# DINOSAURS 101:

## What Everyone Should Know about Dinosaur Anatomy, Ecology, Evolution, and More

W. Scott Persons IV and Philip J. Currie with
Victoria Arbour, Matthew Vavrek, Eva Koppelhus,
and Jessica Edwards

Copyright © 2019 W. Scott Persons IV, Philip J. Currie, Victoria Arbour, Matthew Vavrek, Eva Koppelhus, and Jessica Edwards

All rights reserved. No part of this book may be reproduced, stored in a retrieval system, or transmitted, in any form or by any means (electronic, mechanical, photocopying, recording, or otherwise), without prior written permission from the publisher, except in the case of brief quotations in articles or reviews.

Published by Van Rye Publishing, LLC
www.vanryepublishing.com

Library of Congress Control Number: 2019933340
ISBN-13: 978-0-9982893-4-2
ISBN-10: 0-9982893-4-5

# Contents

# Introduction

IN CONTEMPLATING THE HISTORY of life, Charles Darwin famously wrote: "... from so simple a beginning, endless forms most beautiful and most wonderful have been, and are being, evolved." Since it first began on our planet over 3.5 billion years ago, evolution has gone on to produce intricate microbial microcosms, vast teaming jungles, colonizers of volcanic vents, high-flyers, savage killers, and intelligent problem solvers. It seems a shame that Darwin did not live to see and revel in our current state of biological understanding. He knew of only a fraction of the forms of life that have now been discovered. It is a particular shame that by Darwin's death in 1882, science had only just begun to piece together the fossil skeletons and the true nature of the creatures we call dinosaurs. Because, of all the forms of life that have been and are being evolved, dinosaurs must surely rank among the grandest and the most wonderful.

Paleontology has been called a "gateway" science. You don't have to be a trained scientist to marvel at the skeleton of a *Tyrannosaurus* and to want to know more about its life, evolution, and extinction. Like many people, and perhaps like yourself, I became fascinated with the wonder of dinosaurs when I was very young (at the age of two-and-a-half, or so I am told). But regardless of our age, prehistoric beasts have a special power over our imaginations. As a scientific field, the study of prehistoric life is perhaps unmatched in its ability to tug at our curiosity. If fostered, this curiosity can grow and lead to a fascination with deep scientific questions and with the scientific process itself. That is why this book has been written. I hope that it provides a gateway for those of you who are curious about dinosaurs and that it leads to a greater appreciation of their exceptionalism and a greater understanding of the science of paleontology and the scientific process.

Over 700 species of dinosaurs are now known, and new species are being discovered at a faster rate than ever before, so there has never been a better time to study dinosaur paleontology. More paleontologists than ever before are out looking for dinosaur fossils and studying them back at their museums and laboratories. Half a century ago, most of the known dinosaurs came only from Europe and North America, as did most paleontologists. That is no longer the case.

The fossil beds of China and Mongolia are among the most prolific dinosaur dig sites in the world. Researchers in the badlands of Argentina, Chile, and Brazil have unearthed entirely new kinds of dinosaurs, including the largest species yet known. And work is ramping up in northern Africa and in Australia that promises to greatly expand the known record of dinosaurs. Even Antarctica is now a place of repeat dinosaur discovery. This international work is critical to our understanding of dinosaurs, because dinosaurs were a globally-distributed animal group, with unique forms evolving over time throughout the varied environments of the prehistoric world.

So, there has also never been a more *exciting* time to study dinosaurs. As our understanding of them has improved, dinosaurs have only become more wondrous. Long gone are the assumptions that dinosaurs were simply an "overgrown" kind of reptilian monster—a primitive form of animal life destined for extinction and failure. Today, we understand dinosaurs to have been endothermic (warm-blooded) animals, many covered with feathery fuzz. Many dinosaurs were social animals that lived together in vast groups and that lived with and cared for their young. Some were small animals that burrowed in the ground and climbed through the trees.

We now understand that dinosaurs were (and are) great successes in two important regards. First, there are more species of living dinosaurs than there are mammals. Birds are a branch of the dinosaur family tree that survives and thrives to this day. Second, the reign of the non-avian (non-bird) dinosaurs lasted for over 160 million years. During that time, dinosaurs were the only animals evolving to fill the ecological roles of big land herbivores and carnivores. Over the course of the current "Age of Mammals," our own mammalian group has been evolving to fill those roles for a mere 66 million years. In that regard, dinosaurs have a good 90+ million years of evolution over today's big land animals. In *Dinosaurs 101*, you will learn not to think of dinosaurs as primitive monsters, but as extremely successful ancient animals that were highly varied and surprisingly advanced.

Dr. W. Scott Persons IV
Dinosaur Paleontologist
University of Alberta

CHAPTER 1

# The Skeleton

**LEARNING OBJECTIVE FOR CHAPTER 1:** Understand the diversity of dinosaur anatomy, including bony and soft tissue structures, and identify unique features of the major dinosaur groups.

- **Learning Objective 1.1:** Understand how the sizes of dinosaurs compare to modern animals.
- **Learning Objective 1.2:** Identify the major bones in the tetrapod skeleton.
- **Learning Objective 1.3:** Recognize the two major types of hips in dinosaurs.
- **Learning Objective 1.4:** Identify the skeletal characteristics of the major dinosaur groups.
- **Learning Objective 1.5:** Compare major integumentary types in different dinosaur groups.
- **Learning Objective 1.6:** Recognize muscle insertion points on bones.

# OWEN'S "TERRIBLE LIZARDS"

Welcome to *Dinosaurs 101*! In this book, you will explore the anatomy, ecology, and evolution of one of the grandest and most fascinating animal groups to ever walk the earth.

The term "dinosaur" was invented over 170 years ago by British naturalist Sir Richard Owen. Although Owen named the group, he only had a faint idea of what dinosaurs were really like. At that time, no complete dinosaur skeletons had been found. Only a few fragmentary specimens from a small number of different species were known. There was a jaw, a partial hip, and a few other bits and pieces from the large carnivore *Megalosaurus*; teeth, vertebrae, and limb bones from the herbivore *Iguanodon*; and some ribs, incomplete shoulder girdles, a small portion of skull, and armored plates from the herbivore *Hylaeosaurus*.

**Figure 1.1.** Sir Richard Owen. (Figure in the public domain)

Each of these three dinosaurs had been previously described in the scientific literature, and each had been identified as some form of extinct giant reptile. But Sir Richard Owen was the first to realize that the three shared an unusual combination of anatomical traits that suggested they were all more closely related to each other than any of them were to any living reptile. Among the traits that Owen realized the trio of animals shared were teeth that grew in sockets (like modern crocodiles) and erect limbs (like mammals and birds). These shared similarities, Owen reasoned, could not simply be coincidental, and he put forward the hypothesis that *Megalosaurus*, *Iguanodon*, and *Hylaeosaurus* belonged together in a single natural group. He named that group the "Dinosauria"—"dino" meaning "fearfully great or terrible" and "sauria" meaning "lizards" or "reptiles."

A great deal has changed since Owen coined the term "Dinosauria." Our understanding of what makes an animal a dinosaur has been refined, and the list of shared anatomical features that unites the Dinosauria has lengthened and improved.

## DINOSAUR SIZE

Dinosaurs are famous for being big. Some dinosaurs were the biggest land animals in our planet's history. Today, the only animals that match the size of many sauropods, or "long-necked dinosaurs," are whales (which rely on the buoyancy of water to support their weight). Many other groups of dinosaurs also include huge species, such as the famous *Tyrannosaurus* and *Triceratops*, which both grew larger than elephants.

Although the most well-known dinosaurs are giants, dinosaurs actually came in a wide variety of sizes. Birds belong to a specialized branch of the dinosaur family tree, so, technically, hummingbirds are the smallest dinosaurs. But there were also many varieties of tiny dinosaurs that were only distantly related to birds. For example, the dinosaurs *Microraptor* and *Fruitadens* were less than 1 meter (3.28 feet) long as adults and probably weighed no more than a few pounds.

Big dinosaurs had skeletons with big bones, which tend to be more durable and are also easier for fossil hunters to find. Small dinosaurs had delicate small skeletons that are easily destroyed, and those that survive are not so easily discovered. In this way, the fossil record has been biased toward large dinosaurs. But fossil hunting techniques have improved, and paleontologists have started looking closely at rare fossil beds where tiny dinosaur specimens are preserved, such as those of Liaoning China (see the "Integument" section of this chapter). One of the biggest changes in our understanding of dinosaurs is a new appreciation for just how many small species there were. These small dinosaurs fit into the prehistoric ecosystems in different ways than large dinosaurs did. For instance, we now know there were many small dinosaurs that lived in trees and some that dug burrows underground.

The sizes of some dinosaurs have been inflated by Hollywood and pop culture. For example, the predatory *Velociraptor* was made famous by the *Jurassic Park* franchise, in which it is depicted as an imposing lion-sized terror. In reality, *Velociraptor* was about the size of a coyote—much smaller than an adult human.

## FOSSILS

The study of dinosaurs is a subdivision of the branch of science known as paleontology. **Paleontology** is the study of all prehistoric life. A paleontologist's knowledge of prehistoric life comes primarily from fossils.

A **fossil** is any preserved evidence left behind by a prehistoric organism. The word fossil literally means "dug up," and fossils are usually objects or structures found buried in ancient rock formations. Dinosaur fossils include footprints, eggshells, coprolites (fossil poop), and in rare instances even skin and feather impressions. However, most dinosaur fossils are bones. Bones are partially made of minerals, which do not decay as easily as flesh and other soft tissues. For this reason, bones have a much greater chance of being preserved as fossils, and a dinosaur paleontologist needs to know a great deal about bones.

## THE VERTEBRATE SKELETON

**Adaptations** are traits that have evolved to serve specific functions. Bones are adaptations that help animals to survive by serving four major functions. First, bones passively resist gravity and maintain an animal's form. When you stand up straight, the bones in your legs act like support columns. Your leg bones support your weight, without your muscles needing to actively flex and expend energy.

Second, bones provide a rigid framework for muscle attachment. Raise your right hand high over your head. When you do, you can feel muscles in your shoulder flexing. The bones in your shoulder girdle provide a solid anchor against which your shoulder muscles can pull, and long bones in your arm allow it to move as a single stiff unit.

Third, bones provide protection and can also be major components of horns and other robust weapons. For example, your skull bones form a natural helmet that protects your brain—a delicate organ that could be seriously damaged by an impact with an unexpectedly low doorway or rogue baseball.

Finally, bones store mineral reserves. Often, the resources that an animal needs to grow and develop are plentiful at one time and rare at another. During times of plenty, animals may store a valuable mineral resource, such as calcium, by growing a new bone deposit or by increasing the density of already existing bone. Later, during a time when the resource is scarce, the animal may gain access to stored minerals by reabsorbing some of its bone.

Dinosaurs belong to a group of animals known as vertebrates (and so do you). **Vertebrates** are animals that have two special kinds of skeletal adaptations: skulls and vertebrae. **Vertebrae** are structures made primarily of bone and/or cartilage that surround a portion of the spinal nerve cord. Vertebrae interlock with each other in a series and form the **vertebral column**. Fish, amphibians, turtles, snakes, birds, and mammals are all examples of vertebrates. The first vertebrates were aquatic animals that evolved over 500 million years ago.

Animals that lack vertebrae are called **invertebrates** and include animals like insects, spiders, snails, squids, clams, jellyfish, and worms. Since the origin of animal life, there have always been many more species of invertebrates than vertebrates. However, vertebrates are more numerous when it comes to species of large animals. This success is probably related to the vertebral column's ability to passively support weight and to anchor enlarged muscles.

## Skulls and Jaws

The skull is not a single bone. Rather, the skull is made up of many bones that are tightly locked together. More than any other part of the skeleton, a skull can give a paleontologist great insight into a dinosaur's life. The upper and lower jaws may contain teeth and/or include a beak, and they are critical for interpreting what a dinosaur was adapted to eat. The rear portion of the skull includes the braincase. The **braincase** is a hollow chamber formed by multiple skull bones that houses the brain. There are many small openings into the braincase. Nerves pass through these openings and connect to the brain. The size and shape of a braincase can indicate the size and shape of the brain that it housed, and, therefore, can provide clues to a dinosaur's mental capabilities.

Dinosaur skulls also have multiple pairs of large openings. The **nares** (singular: naris) are the pair of openings for the nostrils. The **orbits** are the pair of openings for the eyes. In some animals, like turtles, there are no other large skull openings, but dinosaurs have several. These additional skull openings are called **fenestrae**. The word "fenestrae" (singular: fenestra) is Latin for "windows."

Behind each orbit, dinosaurs have two fenestrae: the fenestrae on the lateral sides of the skull are called the **laterotemporal fenestrae**, and the fenestrae on the top of the skull are called the **supratemporal fenestrae**. Both the laterotemporal fenestrae and the supratemporal fenestrae provide extra room for large jaw muscles. Between each orbit and naris, dinosaurs have a third fenestrae pair, called the **antorbital fenestrae**. The function of the antorbital fenestrae is unclear. They may have simply been adaptations that made dinosaur skulls lighter, or large sinus cavities housed in the fenestrae may have helped warm the air that dinosaurs breathed.

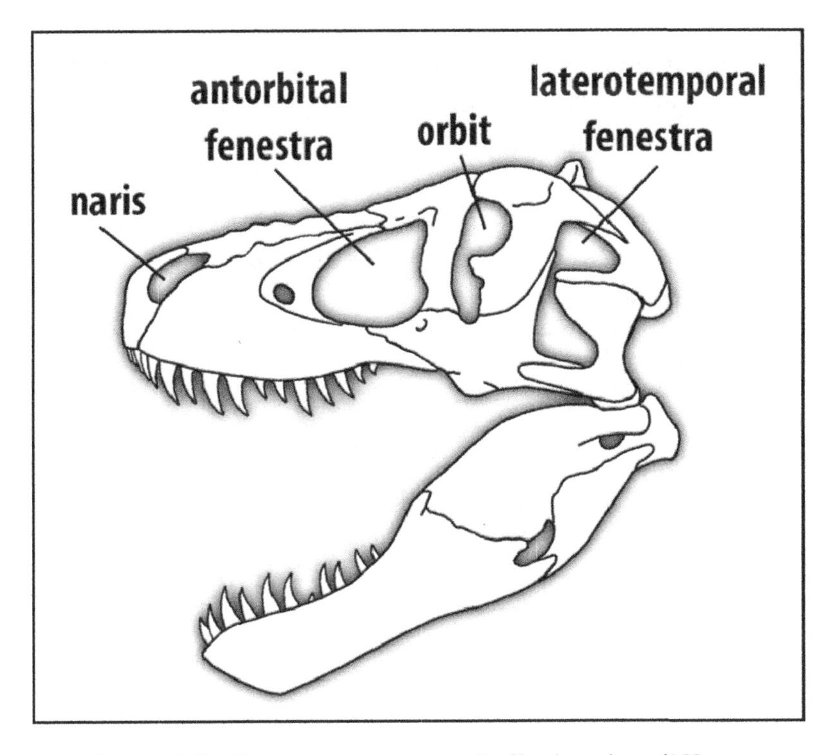


**Figure 1.2.** *Tyrannosaurus rex* skull, showing different openings. (Figure by Veronica Krawcewicz)

## The Axial Skeleton

The vertebral column, or spinal column, is comprised of a series of interlocking vertebrae (singular: vertebra) that begins with the first vertebra in the neck and ends with the last vertebra in the tail. Nearly all vertebrae share a basic form. A vertebra has a spool- or disk-shaped body, called the **centrum**. Above the centrum is the **neural arch**, which covers the neural canal. The **neural canal** is the opening in each vertebra, through which the spinal nerves run. A vertebra may also have processes extending from the centrum or neural arch. **Vertebral processes** provide attachment surfaces for muscles and sometimes provide articulation surfaces for ribs. Two common types of vertebral processes are **transverse processes**, which extend from the lateral sides of the vertebrae, and **spinous processes**, which extend upward from the neural arch.

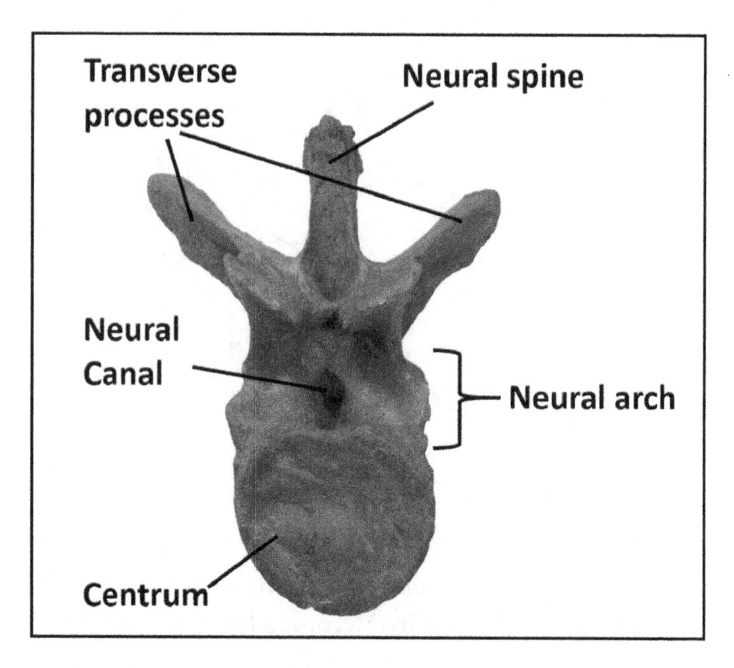


**Figure 1.3.** Vertebra of *Albertosaurus*, in anterior (front) view. (Figure by W. Scott Persons)

Throughout the vertebral column of any animal, the shapes of individual vertebrae vary. In many animals, like most fish, this variation in vertebral shape is slight. However, in animals like dinosaurs and mammals, vertebrae in different regions of the vertebral column have strikingly different shapes.

Vertebrae in the neck are called **cervical vertebrae**. Cervical vertebrae often have extra-large openings for blood and nerve channels and are adapted to support the weight of an animal's head. Vertebrae in the back are called **dorsal vertebrae**. Dorsal vertebrae often have tall spinous processes and large rib articulation surfaces. Vertebrae in the hips are called **sacral vertebrae**. Because the pelvic bones serve as solid anchors for powerful leg muscles, the pelvic bones (later discussed in detail) are fused to the sacral vertebrae. To further increase the strength of the hips, the sacral vertebrae are also fused with one another and form a single solid bone structure called the **sacrum**. Finally, vertebrae in the tail are called **caudal vertebrae**. Underneath caudal vertebrae are bones called **chevrons**. Chevrons protect a large blood and nerve channel and provide support for tail muscles.

In dinosaurs, cervical, dorsal, sacral, and caudal vertebrae may all support ribs (although, in the tail, ribs are usually only present at the base and are tightly fused to vertebrae). The largest ribs are those that connect to the dorsal vertebrae and form the ribcage. In dinosaurs, all dorsal vertebrae connect with ribs; however, in mammals, the dorsal vertebrae close to the hips do not. Also unlike mammals, some dinosaurs had gastralia, or "belly ribs." **Gastralia** are small ribs positioned across a dinosaur's underbelly, underneath the ribcage.

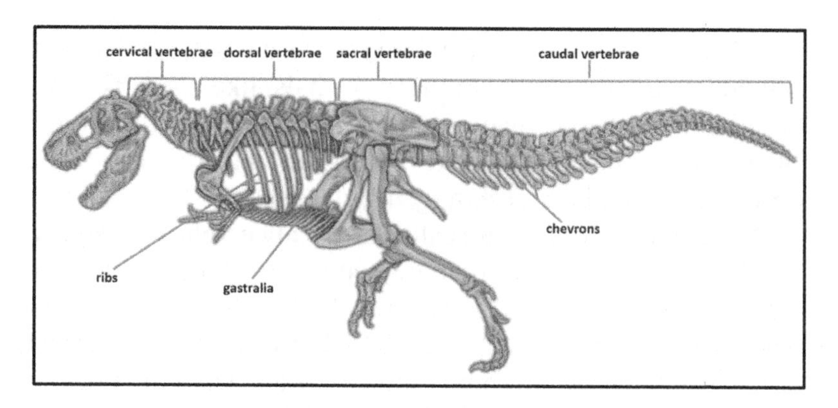


**Figure 1.4.** The axial skeleton of *Tyrannosaurus*. (Figure by Veronica Krawcewicz)

## The Appendicular Skeleton

Dinosaurs, mammals, reptiles, and amphibians all belong to a special group of vertebrates known as tetrapods. "Tetrapod" means "four feet." **Tetrapods** are animals that evolved from an ancient ancestor with four feet and four limbs. Most tetrapods still have four feet and limbs, although some, like humans, have hands instead of front feet, and some, like snakes, have lost their limbs altogether.

The limbs of a tetrapod are connected to the rest of the skeleton by **limb girdles**. The forelimbs connect to the **pectoral girdle**, also called the shoulder girdle. The **scapula**, or shoulder blade, is the largest bone in each side of the pectoral girdle. The hindlimbs connect to the **pelvic girdle**, or hip bones. Each side of the pelvic girdle is composed of three bones that are tightly connected to one another. The upper hip bone is called the **ilium**. It is to the ilium that the sacral vertebrae are fused. Below the ilium are the **pubis** and the **ischium**. The pubis is positioned in front of the ischium, nearer the belly, and the ischium is positioned behind the pubis, nearer the tail. The **acetabulum** is the depression or (as in dinosaurs) the hole in the pelvic girdle into which a hindlimb articulates.

Between the shoulder and elbow is the largest bone in the forelimb, called the **humerus**. Between the elbow and the wrist are two parallel bones, called the **radius** and **ulna**. In most tetrapods, the radius is the thinner of the two. The bones in the wrist are called **carpals**. The bones between the wrist and fingers are called **metacarpals**. Finger bones are called **phalanges**.

The arrangement of bones in the hindlimbs is very similar to that in the forelimbs. Between the hip and knee is the largest bone in the hindlimbs, called the **femur**. Between the knee and the ankle are two parallel bones, called the **fibula** and **tibia**. The fibula forms the shin and is usually the thinner of the two. The bones in the ankle are called **tarsals**. The bones between the ankle and toes are called **metatarsals**. Finally, the bones in the toes are called phalanges (the same name as the bones in the fingers).

The same pattern of bones in the limbs is shared by nearly all tetrapods. Changes in the proportions of the limbs, or in the proportions of particular limb bones, or in limb posture can have a major impact on how a tetrapod moves. For instance, when we humans stand and walk, our heels touch the ground. When dinosaurs stood and walked, only their toes touched the ground. So, the metatarsals of dinosaurs (which are located in the flat of human feet) were tilted upward and contributed to the length of a dinosaur's leg. This helped dinosaurs take longer steps and probably allowed many species of dinosaurs to run much faster than humans can.

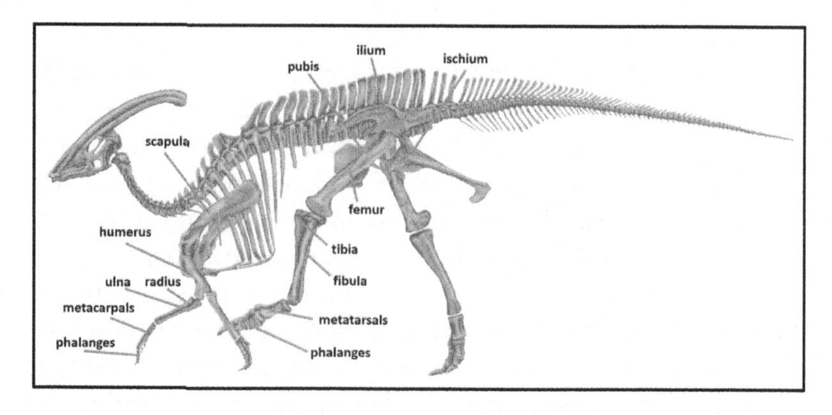


**Figure 1.5.** The appendicular skeleton of *Parasaurolophus*. (Figure by Veronica Krawcewicz)

# BRANCHES OF THE DINOSAUR FAMILY TREES

Skeletal differences and similarities are used to sort dinosaurs into related groups. There are three major groups of dinosaurs: ornithischians, sauropodomorphs, and theropods. Within these major groups, there are many smaller groups.

## Ornithischians

Ornithischian dinosaurs are those that share an evolutionary ancestor that had both a special beak-forming bone in the lower jaw (called the predentary) and a pubis that extended downward and backward toward the tail. The backward-extending pubis was an adaptation that created more space in the ribcage. This extra space was probably filled with extra-large digestive organs. Plant matter is much harder to digest than meat, and generally, herbivores need larger stomachs and intestines than do carnivores. All known ornithischian dinosaurs are thought to have been primarily herbivorous. The beaks, which all ornithischians possess, are also herbivorous adaptations that helped ornithischians chop off large mouthfuls of vegetation. There are five major groups within the ornithischians: heterodontosaurs, thyreophorans, pachycephalosaurs, ceratopsians, and ornithopods.

**Heterodontosaurs** are a rare group of ornithischians. They are small (typically less than 1 meter [3.28 feet] long) and were among the first ornithischians to evolve. Heterodontosaurs stood and ran on long back legs, and their hands each bore three hooked claws. In most dinosaur jaws, all the teeth have a very similar shape, but heterodontosaurs are a notable exception. Many heterodontosaurs have large canine-like tusks in the front of their mouths, just behind the beak.

**Thyreophorans** are better known as the "armored dinosaurs," because the group includes many dinosaurs with bodies covered by osteoderms. **Osteoderms** are bones that develop within the skin and are a common component of animal armor. Today, osteoderms can be observed forming the shells of armadillos and the back scutes of crocodiles. There were two major kinds of thyreophorans: stegosaurs and ankylosaurs.

**Stegosaurs** are quadrupedal thyreophorans, with rows of projecting osteoderm plates down their backs and long osteoderm spikes on their tails. Some stegosaurs also have osteoderm spikes on their backs and over their shoulders. Stegosaur front limbs are much shorter than their hindlimbs. They were not fast runners but could probably pivot quickly and could rear up and stand on their hind legs. Stegosaur heads are small relative to their bodies, and their snouts are narrow.

**Figure 1.6.** The stegosaur *Stegosaurus*. (Figure by Joy Ang and Veronica Krawcewicz)

**Ankylosaurs** are the most heavily armored of all dinosaurs. Ankylosaurs are quadrupedal with short legs and wide ribcages. The necks, backs, flanks, and tails of most ankylosaurs are covered in spiky protective osteoderms. Some ankylosaurs also have large osteoderms on the ends of their tails, forming a mace or "tail club." Unlike their relatives the stegosaurs, most ankylosaurs have robust skulls with thick armor plates.

**Figure 1.7.** The ankylosaur *Anodontosaurus*. (Figure by Joy Ang)

**Pachycephalosaurs** were bipedal with short arms, unusually stout and strong tails, and armored skulls. Some pachycephalosaurs have thick domed skull roofs and backward-pointing horns. The function of pachycephalosaur skull armor will later be discussed in detail, but it has long been speculated that pachycephalosaurs may have rammed predators or have butted heads with each other in competitions for territory or mating rights. In the front of their mouths, behind their beaks, pachycephalosaurs have sharp conical teeth and leaf-shaped teeth in the rear. These front teeth have led some paleontologists to hypothesize that pachycephalosaurs might have been omnivores (adapted to eat meat as well as plants).

**Figure 1.8.** The pachycephalosaur *Pachycephalosaurus*. (Figure by Joy Ang)

**Ceratopsians** are a group that evolved late in the history of dinosaurs. Ceratopsian skulls are expanded in the rear. In most ceratopsians, this rear skull expansion is taken to an extreme, and a large bony frill, or neck shield, is present. Ceratopsians have large parrot-like beaks, and most have dense tightly-packed rows of small teeth in the rear of their mouths. Together, these small teeth form large chewing surfaces and are collectively referred to as **dental batteries**. Many ceratopsian skulls also have large horns. *Triceratops* is easily the most famous of the ceratopsians and is one of the largest. Most large ceratopsians were quadrupedal and have short tails.

**Figure 1.9.** The ceratopsian *Triceratops*. (Figure by Joy Ang)

**Ornithopods** include a wide range of dinosaurs that lack armor and that either walked bipedally all the time or assumed a bipedal stance when running. Many ornithopods are small antelope-sized dinosaurs, but some, like the iguanodonts and hadrosaurs, grew to be multi-ton giants. **Iguanodonts** are large ornithopods with a spike-shaped claw on each hand.

**Figure 1.10.** The iguanodont *Iguanodon*. (Figure by Joy Ang)

**Hadrosaurs** are the famous "duckbilled dinosaurs." Hadrosaurs are ornithopods that evolved late in the history of dinosaurs but were highly successful. Some hadrosaurs have elaborate bony crests, and all hadrosaurs have strikingly large beaks in the front of their mouths and dental batteries in the back.

**Figure 1.11.** The hadrosaur *Corythosaurus*. (Figure by Joy Ang)

## Sauropodomorphs

Sauropodomorph dinosaurs were large herbivores with relatively small heads and elongated necks, which helped them to feed on high-growing vegetation. Among the sauropodomorphs, there were prosauropods and sauropods.

**Prosauropods** were an early group of sauropodomorphs and were the first large-bodied herbivorous dinosaurs to evolve.

**Figure 1.12.** The prosauropod *Massospondylus*. (Figure by Joy Ang)

**Sauropods** were a later group of sauropodomorphs. Many sauropods attained truly gigantic size, and the group includes the largest animals to ever walk the earth. Sauropods stood on four robust and column-like legs. Sauropod vertebrae (particularly the cervical vertebrae) are filled with complex air sacks, which helped to reduce weight. The teeth of sauropods are usually simple and peg-like.

**Figure 1.13.** The sauropod *Diplodocus*. (Figure by Joy Ang and Veronica Krawcewicz)

## Theropods

Theropods were bipedal dinosaurs that shared a carnivorous ancestor. Many theropods were carnivorous and have serrated blade-like teeth and sharp hooked claws, but some were herbivorous, and a few lack teeth altogether. Birds are a kind of theropod, making theropods the only group of dinosaurs that is not completely extinct.

**Figure 1.14.** The theropod *Gorgosaurus*. (Figure by Joy Ang and Veronica Krawcewicz)

**Figure 1.15.** The theropod *Archaeopteryx*. (Figure by Joy Ang and Veronica Krawcewicz)

Both theropods and sauropods share an evolutionary ancestor with a pubis that extended downward and forward, toward the ribcage. For this reason, sauropodomorphs and theropods were long thought by most paleontologists to be more closely related to each other than to ornithischians and were placed together in a group called the **saurischian** dinosaurs. "Saurischian" means "lizard-hipped," while "ornithischian" means "bird-hipped." These terms were used because, in a lizard's hips, the pubis extends downward and forward, and in a bird's hip, the pubis extends downward and backward. Be careful: "saurischian" and "ornithischian" are basic and very old terms, but they can be confusing. Remember, these groups are based on the pubis shape of a shared ancestor, and the groups were named before it was recognized that birds are living dinosaurs.

While some dinosaurs still have the same pubis shape as their ancestor, others have changed it. For instance, despite being the namesake of the term "ornithischian," birds are not ornithischians—birds are not "bird-hipped" dinosaurs! Birds are part of a special group of saurischian dinosaurs that changed their pubis from extending forward to extending backward (unrelated to the similar hip shape of the ancestor of ornithischian dinosaurs).

Here is another potential complication: saurischians and ornithischians share a common ancestor with each other, and paleontologists are not sure what this ancestor was like—whether it had more ornithischian or saurischian traits, or some mosaic of the two. Today, many paleontologists still think theropods and sauropodomorphs are more closely related to each other and still use the term saurischians. But many other paleontologists do not. Theropods may actually be more closely related to ornithischians. For now, how the three major branches of the dinosaur family tree are related to each other remains a scientific puzzle, and it will probably take some new fossil discoveries to finally solve it.

# INTEGUMENT

As has already been discussed, bones are the most common dinosaur fossils because bones decay less rapidly than do softer tissues. This makes it difficult to know what a dinosaur's **integument** (body covering) was like. However, under exceptional circumstances, softer tissues can be fossilized.

Fossil footprints are natural foot molds that were originally made in soft, fine-grained sediments. Sometimes, a footprint may include more than just the rough outline of a dinosaur's foot and may have impressions of foot scales. Skin impressions from other regions of a dinosaur's body can be preserved if a dinosaur was covered by mud shortly after it died and before its flesh rotted away. Direct fossilization of skin and other soft parts is also possible, but such instances are exceedingly rare.

Dinosaur specimens that include a lot of skin fossils are often called "mummies." The first mummified dinosaurs were hadrosaur specimens, found in Wyoming in 1910. These revealed that hadrosaurs were covered with scales and that scales from different regions of the body often had different shapes. Fossil scales are also known from specimens of theropods, sauropods, ceratopsians, stegosaurs, and ankylosaurs. The scaly skin of dinosaurs has a slightly better chance of being fossilized than would our own skin, because scales are partially made of a substance called keratin. **Keratin** is a tough but flexible material that also composes hair, feathers, fingernails, and the outside of claws, beaks, and horns.

A major breakthrough in the study of dinosaur integument came in 1996, when a small theropod specimen with fossil feathers was discovered in Liaoning, China. The feathers had been preserved because the dinosaur's body was buried suddenly in the bottom of a lake by extremely fine ash from a volcano. Since this discovery, many other dinosaur skeletons with feathers have been found in Liaoning. We now know that lots of small theropods had a covering of simple hair-like feathers, and some, like *Microrap-*

*tor*, had feathered wings. In 2012, feathers were first reported from the large tyrannosauroid *Yutyrannus*. At over 1 ton in weight, *Yutyrannus* is the largest known dinosaur with confirmed feathers.

**Figure 1.16.** Forelimb feathers of the small theropod *Microraptor*, from Liaoning, China. (Figure by W. Scott Persons)

Theropods are not the only dinosaurs with hair-like integument. The little ornithischian *Kulindadromeus* and the heterodontosaur *Tianyulong* are also known to have been partially coated in long fibrous integument. The ceratopsian *Psittacosaurus*, although it was mostly covered in scaly skin, had a brush-like arrangement of bristly fibers on its tail. It is not clear whether or not these ornithischian integuments are related to the simple early feathers seen in theropods. If they are, then hair-like insulating feathers were likely a trait common to the ancestors of all dinosaurs.

**Figure 1.17.** The little ceratopsian *Psittacosaurus* had bristle-like structures on its tail. (Figure by Rachelle Bugeaud and Veronica Krawcewicz)

Recall that osteoderms are bones that develop within the skin, so these bones also count as integumentary structures. Osteoderms comprise the armor covering of many types of modern animals, including armadillos, crocodilians, and some lizards. Among dinosaurs, large osteoderms formed the plates and spikes of stegosaurs and the armor and tail-club ends of ankylosaurs. Some sauropods also have osteoderms, although it has been hypothesized that the osteoderms of sauropods were less important for protection and more important as mineral reserves.

# COLOR

What color dinosaurs were has always been a mystery. Recently, a clever new approach has begun to try to solve this mystery, at least for some feathered dinosaurs. Feather colors are not directly preserved in fossilized feathers; however, studies of modern birds have shown that feather color is influenced by the shape and arrangement of **melanosomes**—pigment cells within a feather. Under microscopic examination, melanosomes can be observed in some fossil feathers, and they give clues to a dinosaur's true colors.

Black and gray colors result from long and narrow melanosomes. Brown and reddish colors come from short and wide melanosomes. White feathers have no melanosomes. Iridescence, or "glossiness," like the shiny black and blue feathers of crows and magpies, corresponds to narrow melanosomes that are aligned in the same direction. Based on analyses of fossil melanosomes, it is thought that the dinosaur *Microraptor* was a glossy black or dark glossy blue, and the dinosaur *Anchiornis* is thought to have been black and white with some reddish-brown on its head.

# MUSCLES

Muscles are sophisticated tissues that produce force by contracting. Understanding dinosaur muscles is critical to understanding how dinosaurs moved. Unfortunately, muscles seldom fossilize. Recall, however, that one of the major functions of bones is to provide attachment surfaces and a rigid framework for muscles. Consequently, the shapes of bones often correspond to particular muscle shapes, and muscles often leave behind scars on the surfaces of bones where they attached. Like the pattern of bones in the skeleton, the overall pattern of muscle arrangement is largely the same across all tetrapods. To understand dinosaur muscles, we can look at the muscles of their closest living relatives: birds and crocodiles. The following is an example.

In crocodiles and many birds, there is a large muscle called the **caudofemoralis**. The caudofemoralis pulls backward on the hind leg and is important for powering birds and crocodilians when they walk and run. The caudofemoralis is anchored to the undersurface of the ilium, to the caudal vertebrae, and to the chevrons. It attaches, via a tendon, to the femur. The femurs of crocodilians have a prominence of bone, called a **trochanter**, where the caudofemoralis muscle attaches. In addition to specially-shaped ilia, caudal vertebrae, and chevrons, dinosaurs also have femurs with these same trochanters. So, we can be sure dinosaurs also had a caudofemoralis.

Based on the size of the various anchor points, we can also say that some dinosaurs, like many theropods and hadrosaurs, had a large caudofemoralis relative to the other proportions of their bodies. This tells us that these dinosaurs were adapted for greater hindlimb power and were probably strong runners.

We can even go a step further and compare the position of the muscle attachments. On most theropods, the trochanter is located high on the femur. A high muscle attachment would have allowed the caudofemoralis to repeatedly retract quickly—a useful adaptation for carnivorous animals that depend on their ability to swing their legs fast when sprinting after prey. On hadrosaurs, the trochanter is located further down on the femur (as it is in most herbivorous dinosaurs). This would have reduced the speed at which the caudofemoralis could have repeatedly retracted, but each retraction would have pulled with greater leverage. The slower but high-leverage leg muscles of hadrosaurs would have been important for an animal that needed to be constantly on the move and slowly grazing from one patch of vegetation to the next.

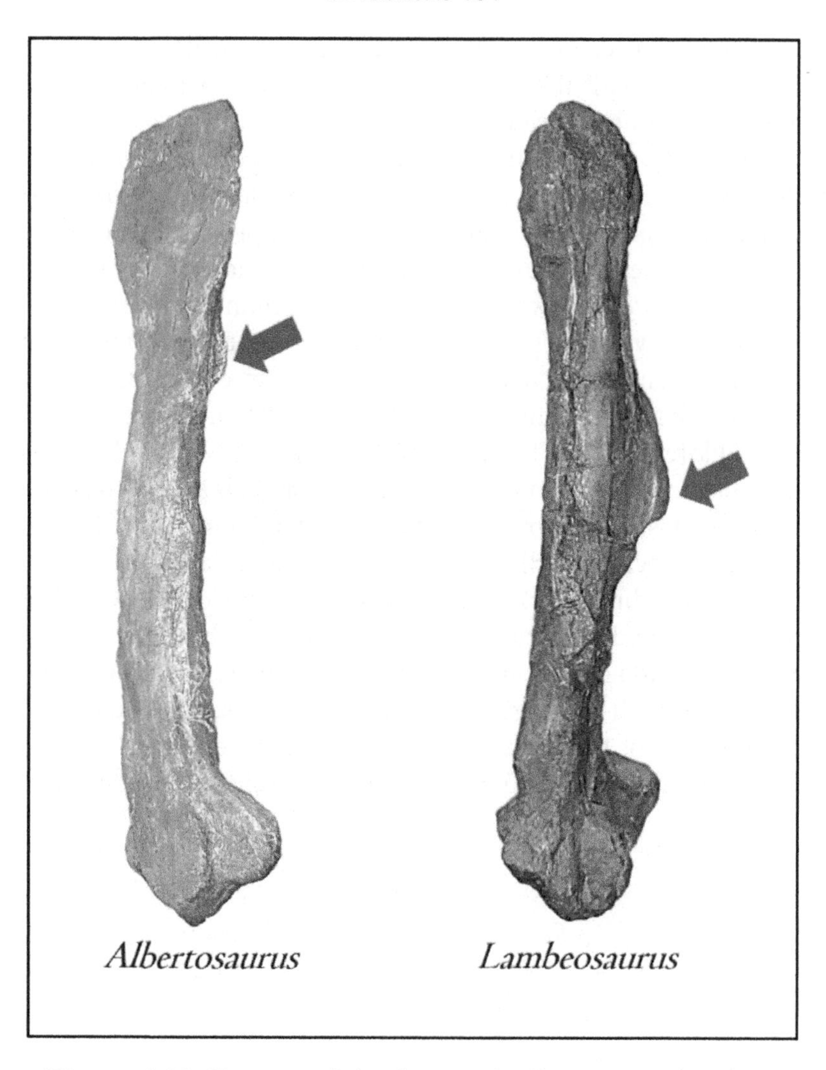


**Figure 1.18.** Femora of the theropod *Albertosaurus* and the hadrosaur *Edmontosaurus*. Arrows point to the trochanter, where the caudofemoralis attaches. (Figure by W. Scott Persons)

CHAPTER 2

# Death and Fossilization

**LEARNING OBJECTIVE FOR CHAPTER 2:** Understand how fossils form and the basic sequence of events in the fossilization process.

- **Learning Objective 2.1:** Classify fossil occurrences as articulated skeletons, associated skeletons, isolated elements, or bone beds.
- **Learning Objective 2.2:** Identify taphonomic features common to dinosaur bones.
- **Learning Objective 2.3:** Define the three main categories of rocks.
- **Learning Objective 2.4:** Identify which kinds of rocks preserve dinosaur fossils.
- **Learning Objective 2.5:** Classify types of fossil preservation.
- **Learning Objective 2.6:** Recognize environments in which fossils can become preserved through burial.
- **Learning Objective 2.7:** Understand what environments are best for preserving fossils.

- **Learning Objective 2.8:** Describe the basic techniques used to collect, prepare, and curate dinosaur fossils.

# AFTER DEATH

We have discussed how fossils can be used to learn about a dinosaur's life. Often, there is more to a fossil's story. The moments immediately after a dinosaur's death may have been an eventful period, and a great deal could happen in the more than 65-million-year interval between a dinosaur's death and the discovery of its fossils. **Taphonomy** is the study of all natural processes that involve an organism after it dies—this includes how it decays, is scavenged by other organisms, becomes fossilized, and erodes.

Although you might think that a dinosaur would naturally stay put after it dies, it is not uncommon for a dinosaur carcass to have been moved a considerable distance from the site of the dinosaur's death. Predators, and later scavengers, may carry carcasses to dens or some other more secure feeding area. Shortly after death, decay may cause a body to swell with putrid gasses, and this may cause the carcasses of even large animals to float easily and be transported by shallow and weakly-flowing water. This phenomenon is known as **bloat-and-float**.

Finding complete dinosaur skeletons is rare. More commonly, only a single bone or a few isolated bones are found. There are many taphonomic factors that can contribute to the disarticulation of a skeleton. Partial consumption by carnivores is one such factor. Carcasses that have rotted for some time may be easily broken apart if swept away by rivers or floodwaters. Water currents may also carry different portions of a skeleton to different locations, based on the weight and shape of the different bones. Prolonged exposure to sunlight gradually weakens and disintegrates bone. Skeletons that become only partially buried will eventually lose their exposed portions. Portions of skeletons may also be trampled by animals or have their mineral content leached away by the roots of plants. These are only a few examples, and there are a large number of other taphonomic factors that can contribute to both the transportation and the **disarticulation** of a skeleton.

Even while buried, taphonomic factors may modify a skeleton. The weight of layers of rock and sediment above a bone may flatten it, and even bone that has already fossilized may be subjected to plastic deformation. **Plastic deformation** occurs when pressure changes the shape of a buried fossil such that, even when the pressure is later removed, the fossil does not return to its original shape. Plastic deformation is an important process to understand and to be mindful of. Otherwise, plastically-deformed fossils may be incorrectly assumed to display their true original shapes.

**Figure 2.1.** The skeleton of a plains zebra (*Equus burchelli*) undergoing modern taphonomic processes. The skeleton has partially disarticulated, and most of the flesh has been stripped away. (Figure by W. Scott Persons)

# FOSSILIZATION ENVIRONMENTS

Fossils may form in a variety of ways. The different ways that fossils form are called **preservation styles**. Most dinosaur bone fossils form through either permineralization or replacement. **Permineralization** occurs when the empty internal spaces of a bone are filled with minerals. These minerals are first dissolved in water and are then deposited in the empty bone spaces as water soaks through the bone. **Replacement** occurs when the original bone gradually decays, and minerals fill the space that the bone once occupied.

To become fossilized, a bone needs to be buried. Burial can occur if an animal dies in its burrow, if it falls into a sinkhole, or if it, or one of its bones, is buried by a predator. But most often, burial occurs when water washes sand or mud over a carcass. Fossilization is, therefore, more common in wet environments than in dry environments where there is no water to help bury carcasses. Fossilization is also more common at low elevations, where sand and mud carried in by water are able to build up, than at high elevations, where sand and mud are often carried away by erosion before they can build up and "permanently" bury and protect a carcass. For this reason, we most often find dinosaur fossils in ancient rocks whose sediments were deposited by rivers, streams, and lakes.

River and stream deposits are called **fluvial deposits**, and lake deposits are called **lacustrine deposits**. Lacustrine deposits have the best chance of preserving soft tissues like hair or feathers. This is because there is relatively little water movement at the bottom of a lake to disrupt the skeleton, and the sediments laid down in lakes are very fine-grained mud—which naturally preserves impressions of soft structures better than coarse sand. Even though there were no marine dinosaurs, dinosaur skeletons are sometimes found in ancient **coastal deposits** and even in deep-water **marine deposits**. Many dinosaurs lived along ancient seashores, and occasionally dinosaurs were washed out to sea by storms and tidal waves.

**Aeolian deposits** form from sediments that are accumulated not by water, but by wind. Aeolian deposits are characteristic of dry deserts. Usually, wind deposits sediments much more slowly than water, making deserts poor places for fossils of large animals to be preserved. For this reason, few desert-dwelling dinosaurs are known. However, one amazing exception is the ancient environment represented by the fossil-rich rocks of the Gobi Desert in Mongolia.

During the Cretaceous Period, much of what is now Mongolia was a sand-swept desert, but it was not all dry. A river also coursed through the desert, and, like the modern Okavango River system of modern Africa, the river formed a large deltaic plain that created a huge oasis. In this deltaic plain, many desert animals, including large dinosaurs, had a chance to be buried by the sediments that were deposited by the river. Dinosaurs in the deserts of Mongolia could also be buried in another way: by sand dunes that suddenly collapsed onto the still-living animal. Sand dune collapses happen when the stability of a dune is compromised by saturating rainstorms. The skeletons of dinosaurs that were buried in this way are often preserved in crouching positions, with their necks bent upward, reaching for air.

## SEDIMENTOLOGY

With only a few rare exceptions, all fossils are found in sedimentary rocks. **Sedimentary rocks** are rocks that form when mineral and organic particles accumulate and become either cemented or compacted together. The two other basic rock types are **igneous rocks**, which form when magma or lava cools, and **metamorphic rocks**, which form deep underground when sedimentary or igneous rocks are changed by extreme heat and pressure.

**Sedimentology** is the science of how sedimentary rocks form. Different kinds of sedimentary rocks form in different environments. Understanding the environmental conditions that led to the formation of the particular sedimentary rocks that contain a fossil can give important clues about the habitat of the fossil organism.

Sedimentary rocks that form from mud and silt are called **mudstone** and **shale**. Lakes are places where large amounts of mud and silt accumulate, and large deposits of mudstone and shale often indicate a former lake-bottom environment. Sedimentary rocks that form from sand are called **sandstone**, and sandstone can indicate a former beach, river channel, or ocean floor environment. **Coal** is a special kind of sedimentary rock that forms from the compressed remains of plants, and coal indicates a former swampy environment. **Limestone** is formed from the accumulation of shells and exoskeletons of small marine invertebrates, and limestone always indicates a former shallow marine environment.

## WHERE TO DIG

Simply because a dinosaur bone managed to beat the odds and become buried and fossilized does not mean that the fossil will ever have a chance to be discovered by a paleontologist. Most of the dinosaur fossils that ever formed have either been destroyed (they have been melted or metamorphosed by geologic processes deep within the earth or have eroded away to dust on the earth's surface) or they remain buried too deep for current excavation technology to detect or to reach. Just as becoming a fossil requires a special set of circumstances, so does becoming a fossil that is discoverable.

To prevent a fossil from eroding away, it must remain buried. However, the burial process must be at least partially reversed in order for the fossil to be near enough to the surface to be found. Dinosaur fossils are, therefore, most commonly found in modern environments where there is considerable recent erosion. Modern environments that are covered with vegetation are bad places to hope to find fossils. Vegetation covers and holds together an environment's topsoil and prevents erosion. **Badlands**, such as those throughout the Canadian and American West, are arid environments where vegetation is sparse, where erosion rates are high, and where large expanses of ancient sedimentary rocks are exposed. Badlands are among the best places to hunt for fossils.

**Figure 2.2.** The badlands of Dinosaur Provincial Park are good places to find fossils, because lots of rocks of the right age and type are exposed at the surface. (Figure by W. Scott Persons)

# EXCAVATION

Using geologic maps, paleontologists can identify locations where there are exposures of sedimentary rocks that are the right age to contain the fossils of dinosaurs. Often, a paleontologist that is hunting for dinosaurs returns to a particular location where fossils have been found before. Whether hunting in a new location or returning to an old location that has previously yielded good specimens, a paleontologist does not simply grab a shovel and immediately commence to digging. First, a paleontologist, and usually an entire paleontological field crew, prospects for promising specimens. The ideal dinosaur skeleton is one that is freshly, and only just barely, exposed above ground. Fossils that are not exposed at all are simply not detectable, and fossils that are completely exposed, and have been for a long time, may be badly weathered.

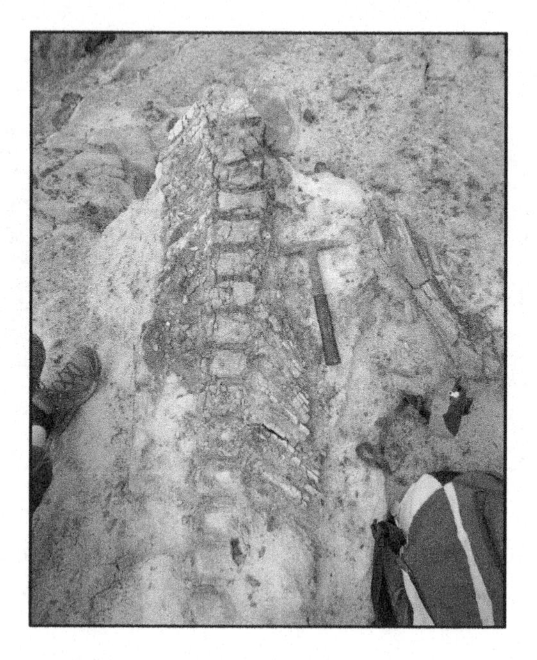

**Figure 2.3.** These are the remains of a hadrosaur tail, in Dinosaur Provincial Park. Once above ground, exposure to the intense summer sun, to rain, and to freeze-thaw cycles in the winter damages the fossil and splinters it into many small shards. (Figure by W. Scott Persons)

Once found, the first step in the excavation of a large fossil specimen is overburden removal. **Overburden** is the rock and earth that covers a fossil specimen and that must be removed before the full extent of the specimen can be judged. Overburden removal usually involves large indelicate tools like shovels, pickaxes, and occasionally even jackhammers and bulldozers. However, such tools are not used in close proximity to fossils. At close distance, the work of the final excavation switches to hand picks and brushes.

Large dinosaur skeletons or bone beds (accumulations of the bones of many dinosaurs) cannot usually be excavated and removed all at one time. Instead, they must be excavated in parts and usually over the course of many field expeditions. Before any one bone is removed, it is important to map its location relative to the other bones. Mapping the relative positions of bones may help in putting a skeleton back together and may also give important taphonomic clues. For instance, if, in a bone bed, all the long limb bones share a similar **orientation**—are observed to lie roughly in parallel with each other—this may indicate that the bones were carried and deposited by a strong river and that this river oriented the limb bones in line with its current.

**Figure 2.4.** Dr. Philip Currie uses a grid square to mark the location of ceratopsian bones on a map. (Figure by W. Scott Persons)

Once a bone has been mapped, it is ready to be dug up. Although fossil bones are mineralized, they are usually brittle and unable to support their own weight. This makes them delicate to transport. To protect a fossil bone on its trip from the field to the laboratory, the bone is wrapped in a layer of protective material (this can be cloth, paper towel, or aluminum foil) and is then covered by strips of burlap that have been soaked in plaster. Once the plaster hardens, it forms a strong and rigid jacket around the fossil. These plaster jackets are not opened until they have reached the laboratory. Then, special glues are applied to the fossils to strengthen them. The final work on removing the rock that surrounds a fossil takes place in the laboratory, and this process often takes more time than the field excavation.

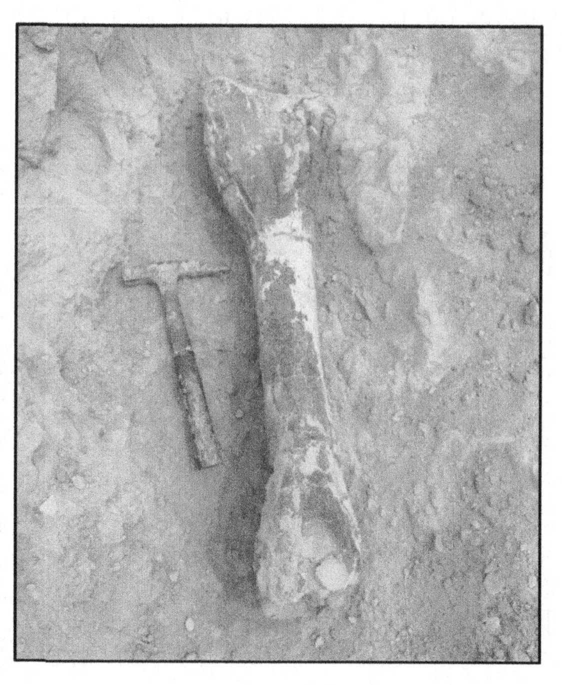

**Figure 2.5.** Stages in the final excavation of a large dinosaur bone: first, the top surface and sides of the fossil are fully unearthed. (Figure by W. Scott Persons)

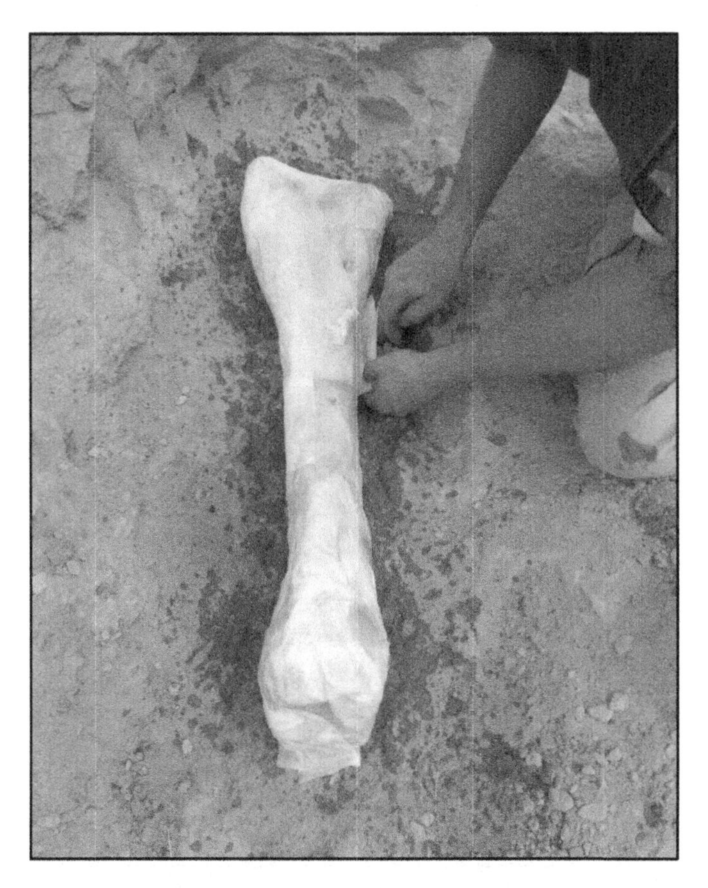

**Figure 2.6.** Stages in the final excavation of a large dinosaur bone: second, a protective and cushioning layer of soft and tight-fitting material is added—in this case, moist paper towels. (Figure by W. Scott Persons)

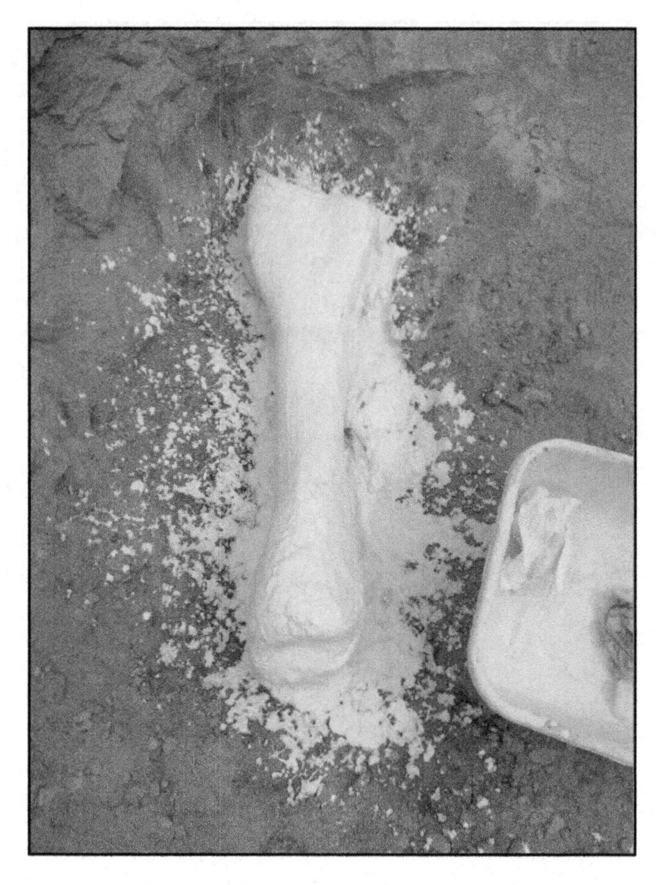

**Figure 2.7.** Stages in the final excavation of a large dinosaur bone: lastly, the fossil is covered by strips of burlap saturated with plaster. Once hardened, the protective jacket is complete, and the fossil may be safely lifted and carried away. (Figure by W. Scott Persons)

CHAPTER 3

# Eating

**LEARNING OBJECTIVE FOR CHAPTER 3:** Understand the feeding habits and feeding adaptations amongst the major dinosaur groups.

- **Learning Objective 3.1:** Recognize the morphological characteristics of a carnivore, herbivore, omnivore, insectivore, and piscivore.
- **Learning Objective 3.2:** Compare tooth replacement in dinosaurs and humans.
- **Learning Objective 3.3:** Understand the characteristics of a dental battery.
- **Learning Objective 3.4:** Recognize dinosaur diet based on dental and other morphological characteristics.
- **Learning Objective 3.5:** Understand the significance of various non-morphological indicators of diet.

# DIET TYPES

The history of non-avian dinosaur evolution spans over 160 million years. In that time, dinosaurs diversified into a variety of forms and became adapted to a variety of ecological roles. The ancestors of dinosaurs were probably carnivores that fed on small reptiles and large insects. Over time, dinosaurs expanded their collective dietary preferences, and different kinds of dinosaurs evolved adaptations to feed on different kinds of food.

Understanding what a dinosaur ate is important if we want to understand how that dinosaur lived and how it fit into a larger ecosystem. Sometimes, a dinosaur skeleton includes fossils of its incompletely-digested last meal inside of its ribcage, but such fossil gut contents are rare. Usually, to figure out a dinosaur's diet, paleontologists must compare its feeding adaptations with those of modern animals whose diets can be directly observed.

**Herbivores** tend to have thin ridged or "leaf-shaped" teeth for shearing and broad flat teeth for grinding. Modern birds lack teeth, but herbivorous birds tend to have short triangular beaks. Herbivores that browse high in trees, but cannot climb, have long legs and necks—like giraffes. **Carnivores** tend to have sharp pointed teeth for piercing, and sharp hooked claws for holding onto struggling prey. Raptorial birds have sharp and hooked beaks and claws.

Like modern carnivores, carnivorous dinosaurs usually have sharp teeth and hooked claws, and, like some carnivorous lizards, most also have teeth with serrated edges. **Serrations** are small sharp bumps on a tooth that are arranged in a line that usually runs from the tip to the base of the tooth. You can see serrations at work on the edge of a steak knife, and, just like the serrated knife edge, serrated tooth edges helped carnivorous dinosaur teeth to slice through flesh.

**Figure 3.1.** The teeth of the carnivorous theropod *Gorgosaurus*. (Figure by Amanda Kelley)

These are only general patterns of adaptation, and some animals with specialized diets are adapted in very different ways. For instance, a parrot is a kind of herbivore known as a frugivore. **Frugivores** eat primarily fruit. The beak of a parrot is sharp and hooked (not unlike the beak of a carnivorous bird) because it needs to rip and tear apart the peels and protective husks of large tropical fruits.

**Figure 3.2.** The hooked beak of a hawk, a carnivore. (Figure by W. Scott Persons)

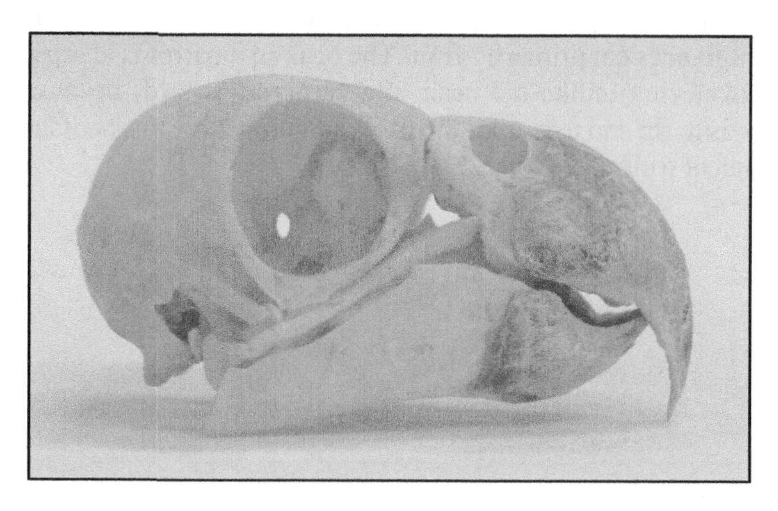

**Figure 3.3.** The hooked beak of a parrot, a frugivore. (Figure by Amanda Kelley)

**Piscivores** are specialized carnivores that primarily eat fish. Piscivores tend to have tall, sharp, conical teeth that usually lack serrations. These adaptations make piscivore teeth good at spearing and holding onto slippery fish. Piscivores also tend to have long jaws that are capable of snapping shut quickly. Piscivorous birds tend to have spear-shaped beaks that are long, straight, and sharp at the tips.

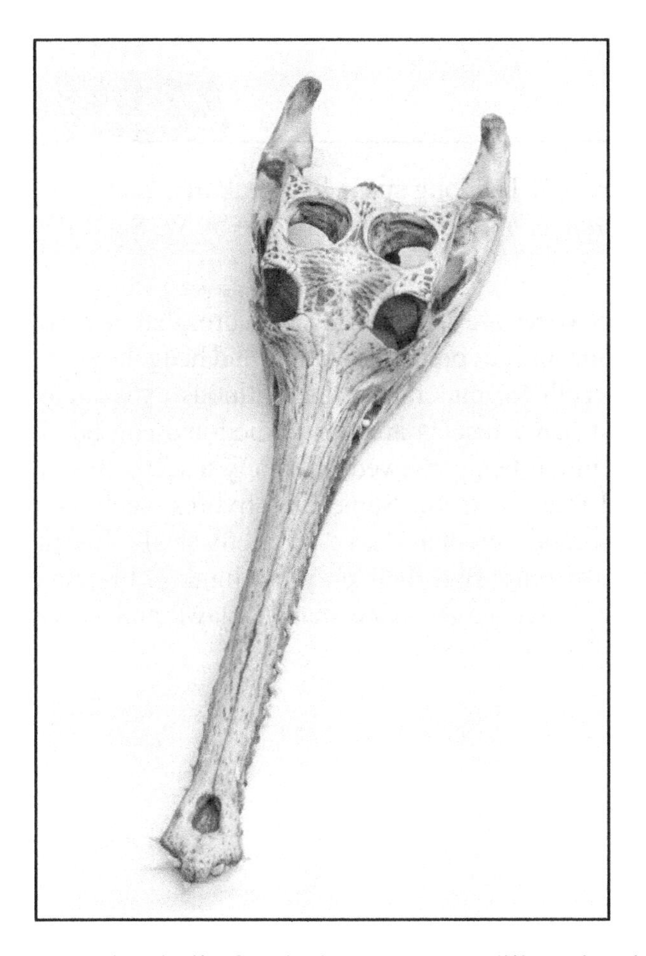

**Figure 3.4.** The skull of a piscivorous crocodilian, the gharial (*Gavialis gangeticus*). (Figure by Amanda Kelley)

**Figure 3.5.** The long spear-like beak of a Common Loon (*Gavia immer*), a piscivore. (Figure by W. Scott Persons)

**Insectivores** are specialized carnivores that primarily eat insects. Some insectivores, like shrews and hedgehogs, have sharp piercing teeth for puncturing the chitinous exoskeletons of insects. But many insects are soft-bodied and can be swallowed whole, without being chewed; so, many insectivores have weak jaws and reduced teeth. Some insectivores, such as anteaters, pangolins, and echidnas, have no teeth at all. Because many insectivores must find their prey by digging, insectivores also commonly have large spade-shaped claws and powerful, but short, limbs.

**Figure 3.6.** The skeleton of a digging insectivore, the echidna (*Tachyglossus aculeatus*). (Figure by Amanda Kelley)

Some carnivores, like hyenas, Tasmanian devils, and alligators, have sharp teeth for puncturing and ripping flesh but also have strong rounded teeth that enable them to crack bones—this is termed **durophagy**. Durophagy also requires extremely powerful jaws.

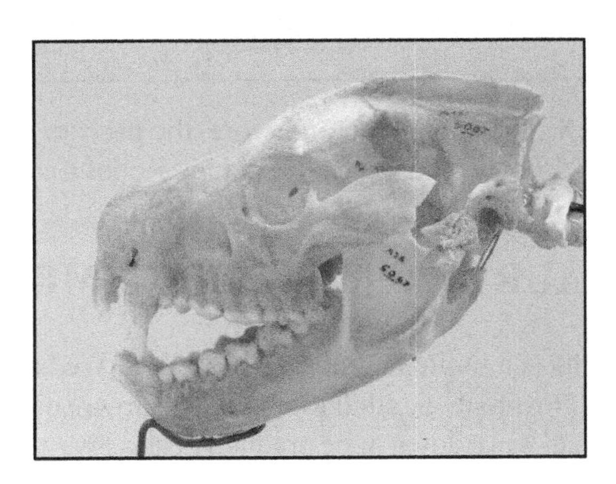

**Figure 3.7.** The skull of a bone-crusher, the Tasmanian devil (*Sarcophilus harrisii*). (Figure by Amanda Kelley)

**Omnivores** are animals that eat significant amounts of both meat and plants. We humans are a good example of an omnivore, as are pigs, most bears, rats, crows, and many turtles. Omnivores tend to have either unspecialized beaks and teeth or a variety of teeth with different shapes (some shaped like those of herbivores and others like those of carnivores). In your mouth, you have pointed canines, which have a shape characteristic of a carnivore, and you also have molars, which have a shape characteristic of an herbivore.

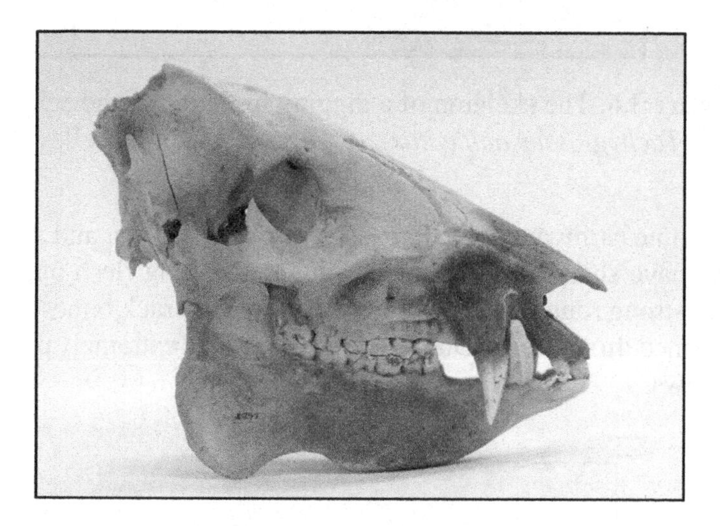

**Figure 3.8.** The skull of an omnivore, the peccary (*Tayassu pecari*). (Figure by Amanda Kelley)

# DINOSAUR TOOTH REPLACEMENT

You and I have a limited number of tooth sets. We have lost our baby teeth. Now that our adult tooth set has grown, our teeth will eventually get worn down or be gradually eaten away by bacteria. If we live long enough, toothlessness or dentures is our inevitable fate.

That is true of most mammals. However, dinosaurs were like modern sharks and crocodiles and were constantly growing new teeth throughout their lives. As new teeth grew in below older teeth, the new teeth would be pushed upward and would eventually replace the old. When an old tooth was ready to be replaced, its root (the portion of the tooth that anchored it in the gums and jaw) would be reabsorbed. **Resorption** is the chemical process by which a dinosaur breaks down its own teeth and bones so that the minerals and nutrients that compose them can be reused. After a new tooth was ready to replace an old one, and after the old tooth's root was reabsorbed, the top, or "crown," of the old tooth could be shed.

Teeth that are ready to be shed still usually require a little help getting free from the little bit of gum that surrounds their base. Remember wiggling and fidgeting with your loose baby teeth? Frequently, loose teeth are shed while an animal is feeding. Throughout their lives, dinosaurs were constantly in the process of replacing teeth. It is estimated that *Tyrannosaurus rex* replaced each tooth once every 1.5 to 2 years. So, shed dinosaur teeth are not uncommon fossils to find and can be easily identified as shed because they are usually well worn and lack roots. Shed teeth can be another useful tool for understanding a carnivorous dinosaur's diet. Often, the skeletons of dinosaurs will be discovered with many shed teeth nearby. That tells us that the dinosaur to whom the shed teeth belong was probably feeding on the other dinosaur's carcass.

**Figure 3.9.** Comparison of a tooth removed from the fossil jaw of a tyrannosaur to a shed tyrannosaur tooth found in association with the disarticulated skeleton of a hadrosaur. (Figure by W. Scott Persons)

# CARNIVOROUS DINOSAUR ADAPTATIONS

**Deinonychosaurs** (popularly known as the "raptor" dinosaurs) are a group of theropods with serrated blade-like teeth and a large sickle-shaped claw on each hind foot. The famous theropod *Velociraptor* is a kind of deinonychosaur. Their special foot claws resemble the retractable claws of modern cats and could be raised off the ground. Keeping their claws raised would have prevented the claws from scratching the ground as deinonychosaurs walked, and this would have kept the claws sharp. Like retractable cat claws, deinonychosaur foot claws are probably specialized weapons used to slash prey and may have also helped some deinonychosaurs climb trees.

**Figure 3.10.** Sickle-shaped foot claws of a deinonychosaur. (Figure by W. Scott Persons)

**Spinosaurs** are a group of theropods with skulls that strongly resemble those of crocodiles. Spinosaurs are thought to be piscivores. Like many modern piscivores, spinosaur teeth are conical, have sharp tips, and have few or no serrations.

**Figure 3.11.** The spinosaur *Spinosaurus*. (Figure by Joy Ang and Veronica Krawcewicz)

**Alvarezsaurs** are a group of small theropods with short front limbs and compact hands. Alvarezsaurs are thought to have been insectivores. Like many modern insectivores, most alvarezsaurs have reduced teeth and short, but strong, front limbs. The alvarezsaur *Shuvuuia* has one large spade-shaped claw on each hand. Its other forelimb claws and fingers were tiny and appear to have been useless.

**Figure 3.12.** The alvarezsaur *Shuvuuia.* (Figure by Joy Ang and Veronica Krawcewicz)

**Tyrannosaurs** are a group of theropods that evolved late in the history of dinosaurs and have reduced front limbs and robust skulls. Tyrannosaur teeth have serrated edges and are well adapted for puncturing and cutting flesh. However, most tyrannosaur teeth have blunt tips, and the attachment sites for jaw muscles in the skulls of tyrannosaurs indicate a capacity for tremendous biting force. It has been estimated that *Tyrannosaurus rex* (the largest of all known tyrannosaurs) had the most powerful bite of any animal (living or extinct). These adaptations indicate that tyrannosaurs may have been capable of durophagy. Note: durophagy is not synonymous with scavenging.

**Figure 3.13.** The tyrannosaur *Tyrannosaurus*. (Figure by Joy Ang and Veronica Krawcewicz)

**Scavenging** refers to the consumption of an already dead animal by a carnivore that did not play a part in killing it. Durophagy can be beneficial to a scavenger because it may allow a carnivore to access nutrients within the bones of a carcass that has already been picked over by other carnivores. However, many durophagous carnivores crush and consume the bones of animals that they themselves have killed, and many animals that are not capable of durophagy regularly scavenge. In fact, vertebrate carnivores that only scavenge are rare, as are those that never scavenge. Scavenging is an opportunistic part of virtually every carnivore's life.

# HERBIVOROUS DINOSAUR ADAPTATIONS

The walls of plant cells are made of a compound called cellulose. **Cellulose** is tough stuff, and it makes plants a difficult source of food. Animals cannot digest cellulose on their own. Animals need help from bacteria that live within their stomach and intestines. Even with the help of bacteria, getting all the raw energy that a large animal needs to survive from plants is not easy. Chewing food before sending it down to the digestive organs helps, because chewing breaks plants into smaller pieces that are easier for bacteria and digestive enzymes to envelope. The dental batteries of some herbivorous dinosaur groups are, therefore, one way of dealing with the challenge of cellulose (and we will discuss dental batteries in a moment). Other herbivorous dinosaur groups evolved different solutions.

Oviraptorosaurs and ornithomimids are two kinds of herbivorous theropods. Many oviraptorosaurs and ornithomimids lack teeth, but some oviraptorosaur and ornithomimid skeletons have small masses of little stones inside their ribcages. These stones were once part of the dinosaurs' gastric mills. A gastric mill is a special stone-filled digesting organ located near the stomach. Many modern birds, including chickens, have a gastric mill, which they fill by swallowing pebbles that they pick up from the ground. Gastric mills help these toothless animals "chew" their food. Swallowed plants are first sent into the gastric mill, where muscular contractions grind the rocks against each other and against the plants. This works just like grinding teeth, and the chewed-up bits of plants then continue into the stomach and are ready to be enveloped by bacteria and digestive enzymes.

**Figure 3.14.** Skeleton of the oviraptorosaur *Caudipteryx*. Arrow points to the mass of stones that filled the gastric mill. (Figure by W. Scott Persons)

Ankylosaurs and sauropods have simple teeth that could be used to nip off vegetation but could only help break down their food a little. What these dinosaurs lacked in chewing ability, they made up for with guts. Ankylosaurs and sauropods have huge ribcages that housed immense digestive organs. Although it would have taken a long time for these dinosaurs to digest plant matter, they still got the energy they needed thanks to their extensive series of digestive vats and the sheer volume of food their digestive tracks were able to hold.

**Figure 3.15.** The skull of the sauropod *Diplodocus*. (Figure by Amanda Kelley)

**Figure 3.16.** The small leaf-shaped tooth of an ankylosaur. (Figure by Amanda Kelley)

## Dental Batteries

Recall that dental batteries are arrangements of densely-packed teeth that collectively form a single large chewing surface, and recall that two groups of dinosaurs evolved dental batteries: hadrosaurs and ceratopsians. Because the individual teeth that make up dental batteries are small and because chewing grinds teeth down quickly, dinosaurs with dental batteries replaced their teeth rapidly. In the skull of a hadrosaur, there can be over 1,000 teeth. Most of these teeth were not actively contributing to the chewing surface. Instead, they are replacements that were already fully formed and waiting in line.

The chewing surfaces of dental batteries are complex. Dinosaur teeth are made of a variety of hard tissues, including enamel (which usually covers the outside of a tooth) and dentine (which is usually common on the inside of a tooth). As a tooth in a dental battery was ground down, different tooth tissues were exposed, and these different tissues would be ground down at a slightly different rate—making the chewing surface slightly uneven. The chewing surface of a dental battery is not simple, uniform, or smooth. It is intricate, varied, and abrasive.

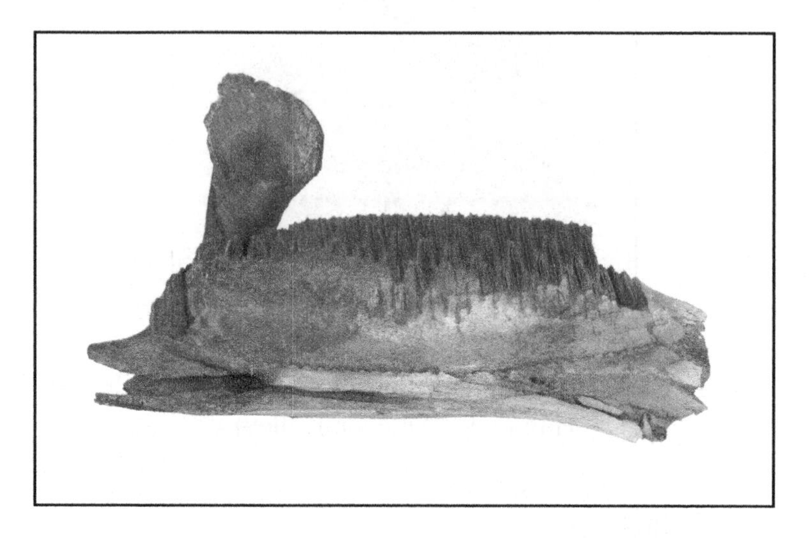

**Figure 3.17.** The lower jaw of a hadrosaur, in medial (tongue-side) view. (Figure by Amanda Kelley)

**Figure 3.18.** The beak and jaws of the ceratopsian *Centrosaurus*. (Figure by Amanda Kelley)

The dental batteries of hadrosaurs and ceratopsians are unrelated (that is, they evolved independently, and hadrosaurs and ceratopsians do not share a common ancestor that possessed dental batteries). The way hadrosaurs and ceratopsians used their dental batteries was also slightly different. In hadrosaurs, the chewing surfaces formed by the dental batteries are almost horizontal. When hadrosaurs chewed, they moved their jaws backward and forward and also from side to side. The chewing surfaces formed by the dental batteries of ceratopsians are almost vertical. Teeth in the jaws of ceratopsians would have slid together like scissor blades, with the opposing lateral sides of the teeth doing most of the grinding. The dental batteries of both hadrosaurs and ceratopsians are inset in the jaw (that is, they are positioned close to the tongue). Inset teeth probably helped make room for large cheeks, and cheeks are important for holding in food while an animal chews.

## OTHER DIET CLUES

Studying adaptations is not the only way to figure out a dinosaur's diet. As mentioned earlier, sometimes dinosaur skeletons included fossil gut contents. Fossil gut contents are termed **cololites**.

Cololites from hadrosaurs and ankylosaurs contain fossil plant material. Two specimens of the wolf-sized theropod *Sinocalliopteryx* contain fossil gut content, and these show that *Sinocalliopteryx* ate birds and small deinonychosaurs. More cololites are known from the deinonychosaur *Microraptor* than from any other theropod. *Microraptor* had a diverse diet and evidently ate small mammals, birds, and fish. A cololite from the spinosaur *Baryonyx* includes fish bones and has helped to support the conclusion that spinosaurs were piscivores (although the cololite also included bones from a large ornithopod; so, spinosaurs ate both fish and other dinosaurs).

Carnivorous dinosaurs often left bite marks on the bones of the dinosaurs they fed on. Tooth mark evidence shows that ceratopsians and hadrosaurs were commonly eaten by tyrannosaurs. Deep puncturing bite marks confirm that tyrannosaurs were capable of durophagy.

One last source of diet information comes from coprolites. **Coprolites** are fossil poop. Although it is often difficult to identify what kind of dinosaur a particular coprolite came from, coprolites can give information not only on what a dinosaur ate but also how it was digested. Coprolites that have been identified as a tyrannosaur's contain large quantities of bone and show not only that tyrannosaurs were durophagous but that the bone tyrannosaurs consumed passed completely through their digestive tracts (unusual even among other durophagous animals).

**Figure 3.19.** Coprolite specimens, in the University of Alberta collection. (Figure by Amanda Kelley)

CHAPTER 4

# Moving Around

**LEARNING OBJECTIVE FOR CHAPTER 4:** Understand the general modes and styles of locomotion in major dinosaur groups.

- **Learning Objective 4.1:** Classify stances in modern animals and dinosaurs.
- **Learning Objective 4.2:** Compare locomotion styles in modern animals.
- **Learning Objective 4.3:** Understand the terms facultative biped, obligate biped, and obligate quadruped.
- **Learning Objective 4.4:** Describe the characteristics of an ichnofossil.
- **Learning Objective 4.5:** Recognize features associated with endothermy and ectothermy.
- **Learning Objective 4.6:** Evaluate the evidence for "warm- or cold-bloodedness" in dinosaurs.

# SPRAWLING AND ERECT STANCES

Lizards, turtles, crocodiles, and salamanders all have what is termed a **sprawling** stance. In a sprawling stance, an animal's humerus and femur project horizontally, with elbows and knees strongly bent. Mammals and birds have what is termed an **erect** stance. In an erect stance, an animal's humerus and femur project vertically, such that all the limbs point straight down from their girdles.

An erect stance has a number of advantages over a sprawling stance. One advantage is that an erect stance positions the limb bones directly under the body. This allows the limb bones to passively support the body's weight without muscles having to strain. Holding a "push-up position" forces our forelimbs into a sprawling stance, and it is hard, because supporting our weight with bent arms requires our muscles to do a lot of work.

Not surprisingly, most animals that have a sprawling posture do not use their limbs to support their weight very often. The life of a lizard is mostly spent resting on its belly. Lizards are relatively inactive and rise to walk and run infrequently. So, a sprawling posture suits lizards just fine. Active animals, like mammals and birds, need a more efficient stance. Most lizards are also not very large, so they have little weight to support. Naturally, the weight-supporting benefit of an erect posture is more helpful to larger animals that have more weight that needs supporting. Another advantage of erect posture is that it allows all the limb bones to fully contribute to the length of a stride. This improves speed, because, if every step you take is longer, you can potentially cover ground more quickly.

All modern tetrapods share an ancestor that had a sprawling stance. Birds and mammals evolved their erect stances independently of each other. Did dinosaurs sprawl like crocodiles and lizards, or stand erect like mammals and birds? Erect and sprawling postures are easy to identify based on footprint trails, limb joints, and the articulation angles of limb girdles. The evidence is clear: dinosaurs stood erect.

**Figure 4.1.** It was once thought that many dinosaurs, like *Diplodocus*, had a sprawling stance. Today, we know that dinosaurs had an erect stance. (Figure by Heinrich Harder, 1916)

## CURSORIAL AND GRAVIPORTAL LIMBS

An erect posture is advantageous for fast locomotion and for supporting body weight. Among animals with an erect posture, there are those that have evolved to maximize these advantages. **Cursorial** limbs are limbs specially-adapted for fast locomotion. To further increase stride length, cursorial limbs are elongated. In particular, cursorial limbs tend to have very long lower leg bones (the bones below the elbows and knees).

Cursorial animals also often stand on their toes—**digitigrade** posture—or stand only on toenails that have been modified into hoofs—**unguligrade** posture. Cheetahs and ostriches are modern examples of animals with cursorial limbs and digitigrade posture. Horses and antelope are modern examples of animals with cursorial limbs and unguligrade posture. We humans are not cursorial, and we stand simultaneously on our toes, the flat of our feet, and our heels—**plantigrade** posture.

**Graviportal** limbs are specially-adapted for supporting extreme body weight. Graviportal limbs have bones that are robust and heavy. Graviportal limbs also tend to have large feet with large fleshy pads. These big feet and pads provide a solid support base and help to absorb impacts when walking. Graviportal limbs tend to be short and, when walking, their joints bend as little as possible. Elephants are modern examples of animals with graviportal limbs.

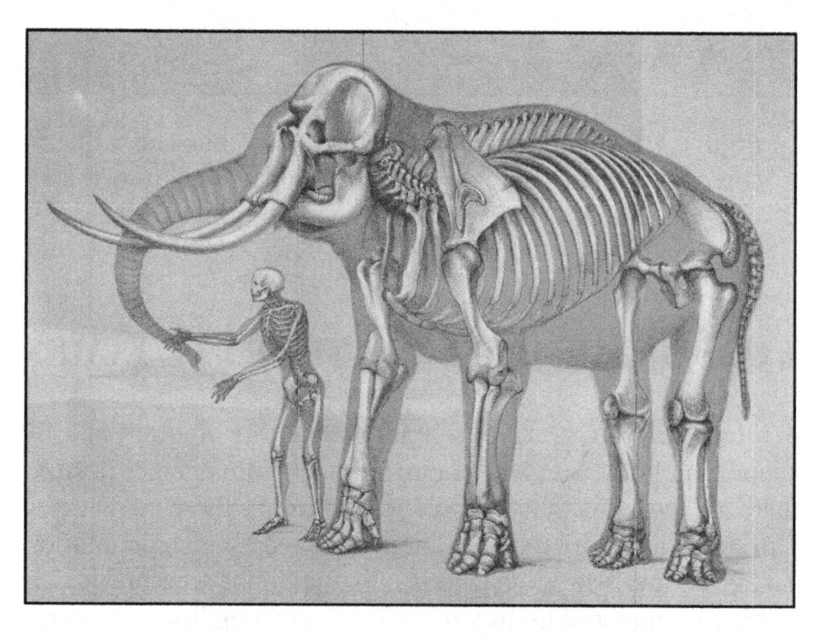

**Figure 4.2.** Compare the robust graviportal limbs of the Asian elephant (*Elephas maximus*) with the more slender bones of the human (*Homo sapiens*). (Figure by Benjamin Waterhouse Hawkins, 1860)

# QUADRUPEDS AND BIPEDS

Animals that only walk and run on two legs, like birds and adult humans, are termed **obligate bipeds**. Animals that only walk and run on four legs, like turtles and horses, are termed **obligate quadrupeds**. Some animals, like basilisk lizards, walk on all four legs but rise on two legs to run. Such animals are termed **facultative bipeds**.

The ancestor of all dinosaurs was an obligate biped. Most dinosaurs remained adapted to carry a majority of their weight on their hind legs and could probably at least stand on only two feet. Nevertheless, we classify sauropods, stegosaurs, and ankylosaurs as obligate quadrupeds, because, even if many of them could stand on two legs, it is unlikely that they frequently attempted to walk bipedally.

Prosauropods are tricky: many were probably bipedal, but whether or not they were obligate or facultative bipeds is not always easy to determine. Some small ceratopsians were obligate and facultative bipeds, and larger ceratopsians were obligate quadrupeds. Pachycephalosaurs and theropods were obligate bipeds, as were most small ornithopods. Our understanding of the postures of large ornithopods, including hadrosaurs and iguanodonts, has changed over the years. Both of these groups have strong hind legs that are significantly longer than their front limbs. This indicated a bipedal stance; however, fossil footprints reveal a different story, as the following section discusses.

# ICHNOFOSSILS

**Ichnofossils** are fossils that record traces of biologic activity. Fossil footprints, tooth marks, and burrows are all examples of ichnofossils. Fossil footprints provide the best direct evidence of how dinosaurs moved. To become fossilized, a footprint must first be made in soft mud. The mud must then dry out and harden. Then, to protect the hardened footprint from erosion, it must be buried but eventually re-exposed so that paleontologists can identify it.

Naturally, the odds of that sequence of events happening to any particular footprint are small. But consider how many footprints one dinosaur could have made throughout the entirety of its life and how many dinosaurs were alive during the more than 160 million years of dinosaur rule. In fact, dinosaur footprints are not uncommon fossils. Often, where one fossil footprint is found so are many others. Sometimes an entire series of dinosaur footprints are found. These fossil footprint assemblages are called **trackways**.

Studies of dinosaur trackways have helped to change our understanding of dinosaur posture and locomotion. For instance, it was once widely imagined that bipedal dinosaurs stood and walked in a way not unlike the movie monster Godzilla—with their belly and torso held vertically above their hips. This posture would have tilted dinosaur tails downward and caused a dinosaur's tail to drag behind it.

**Figure 4.3.** Early depictions often showed iguanodonts and hadrosaurs in Godzillian postures. (Figure by Alice Woodward, 1895)

However, while fossil lizard and crocodile trackways often have tail-drag marks, dinosaur tail-drag marks are rare. We now know that most bipedal dinosaurs held their body in a more horizontal position and that both bipedal and quadrupedal dinosaurs held their tails off the ground.

**Figure 4.4.** Tracks of a modern iguana lizard, with a tail-drag trail down the center. (Figure by W. Scott Persons)

The trackways of hadrosaurs and iguanodonts have deep imprints left by their hind feet and show that these dinosaurs carried most of their weight on their hind legs. However, the trackways of hadrosaurs and iguanodonts also record shallow tracks made by their front feet. Hadrosaurs and iguanodonts were probably facultative bipeds that walked on all fours most of the time but likely reared up on only their back legs to run. The trackways from sauropods reveal large fleshy pads on the back feet. This confirms that sauropods had graviportal limbs.

**Figure 4.5.** Modern conception of *Iguanodon.* (Figure by Joy Ang)

Trackways can also be used to determine how fast dinosaurs moved. When we run, we tend to take long steps, and so do most other animals. As noted previously, longer strides enhance speed. From trackways, we can measure the lengths of dinosaur strides and can usually estimate dinosaur leg lengths from the proportions of their footprints. From these two measurements, it is possible to estimate how fast a dinosaur was moving when its footprints were made. Unfortunately, because fossil footprints must be made in mud and because animals seldom run at full speed when stepping through sticky muck, trackways tell us about dinosaur walking speeds but not usually about dinosaur running speeds.

## ENDOTHERMS AND ECTOTHERMS

In many ways, the limb adaptations of dinosaurs are more similar to those of modern mammals and birds than they are to modern reptiles. Like birds and mammals, dinosaurs were likely "warm-blooded." The terms "**warm-blooded**" and "**cold-blooded**" are antiquated and can be misleading, as the blood of a "warm-blooded" animal is not necessarily any warmer than the blood of a "cold-blooded" animal. To avoid this confusion, we will use the term ectotherm to refer to "cold-blooded" animals and the term endotherm to refer to "warm-blooded" animals.

**Ectotherms** are animals that adjust their internal body temperatures through behaviors that depend on temperature differences within their environment. For instance, to warm up, lizards bask in the sun or on top of hot rocks, and to cool down, lizards seek out shade or cool burrows. **Endotherms** are animals that regulate their own body temperatures through metabolic processes. To warm up, endotherms burn energy to generate internal heat, and to cool down, they may sweat or pant.

Being an endotherm comes at a high cost. In order to maintain a constant optimal body temperature, endotherms must expend large sums of energy. Pound for pound, this means that endotherms must successfully consume a great deal more food than must ectotherms. Due in large part to this significant drawback, most organisms are ectotherms, not endotherms.

However, endotherms do have a few significant advantages. First, endotherms can survive in cold climates, where finding a warm place to absorb heat is impossible. Second, endotherms are always ready for action. On a cold night or morning, before the external environment has had a chance to warm up, endotherms can function the same as they could in the middle of the afternoon. Under the same conditions, ectotherms may be sluggish, making them easy predators to avoid or easy prey to catch. Third, although activities like sunbathing may not waste valuable energy, they do waste valuable time. While ectotherms are forced to spend time basking in the sun or sheltering in the shade, endotherms can be out and about. Although they pay an energy cost, endotherms also do not need to take frequent stops and can maintain high activity levels.

The limbs of dinosaurs, which appear well adapted for the more active lifestyle of endotherms, are one of many arguments supporting the conclusion that dinosaurs were endotherms. Other evidence comes from the discovery that some dinosaurs had simple hair-like feathers. Endotherms benefit from insulating integument to help hold in the body heat that they burn energy to produce. When it comes to being large land-living carnivores and herbivores, endotherms tend to outcompete ectotherms (think about how few large modern reptiles there are compared to mammals). So, the overall pattern of dinosaur ecological success is most consistent with the pattern that would be expected for a group of endotherms.

The bones of dinosaurs also support the conclusion that they were endotherms. **Histology** is the technique of slicing samples of bones into very thin sections, such that the internal structure of the bone can be observed under magnification. Bone cells are called **osteons**. We know from studies of modern animals that endotherms grow their bones more quickly and have their osteons arranged in a different pattern than ectotherms. Dinosaur histology studies show that dinosaur bones grew fast and that dinosaur osteons were arranged like those of endotherms.

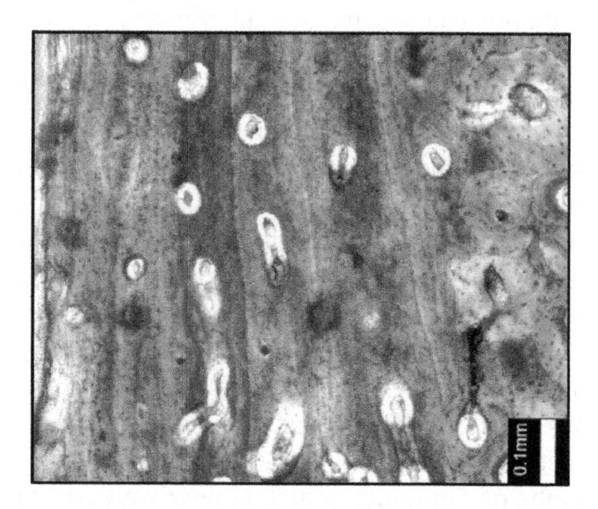


**Figure 4.6.** Thin section of a hadrosaur limb bone. When you cut apart a dinosaur bone and make a thin section of it, you can see the cells that form the bone. (Figure by Mike Burns and Victoria Arbour)

Although it seems clear that many small feathered dinosaurs were endotherms, there is still debate over whether all large dinosaurs were endotherms. It has been suggested that, instead of being endotherms, large dinosaurs were **gigantothermic**. As any shape increases in size, its surface area increases more slowly than its volume. This is called the **square–cube law**. Larger animals, therefore, have relatively less surface area than do smaller animals. It is theorized that, even if big dinosaurs were ectothermic, their low ratio of surface area to volume would have prevented them from losing significant heat to the outside world, and, thus, they could have lived active endothermic-like lives without actually needing to produce body heat by burning energy. However, the theory of gigantothermic dinosaurs remains to be proven and lacks supporting evidence.

CHAPTER 5

# Birth, Growth, and Reproduction

**LEARNING OBJECTIVE FOR CHAPTER 5:** Understand a generalized life history of a dinosaur, from birth through adulthood, including reproduction, and understand the major techniques used to evaluate growth stages and growth rates in dinosaurs.

- **Learning Objective 5.1:** Understand the basic characteristics and evolutionary significance of the amniotic egg.
- **Learning Objective 5.2:** Understand terms related to the gross anatomy and histology of bones.
- **Learning Objective 5.3:** Recognize examples of sexual dimorphism in extant animals and dinosaurs.
- **Learning Objective 5.4:** Understand the evidence for and against parental care in different groups of dinosaurs.

## AMNIOTES

A little over 312 million years ago (long before the evolution of dinosaurs or mammals), a major milestone in tetrapod evolution was reached: the **amniotic egg**. Prior to this adaptation, all tetrapods laid eggs that were similar to those of modern frogs and salamanders and could not retain water. Such eggs would dry out and die if not laid in a wet, humid place. Amniotic eggs are different. They have encapsulating membranes that are watertight. Animals that lay amniotic eggs are called **amniotes**. Being able to hold in their own water, amniotic eggs can be laid in dry habitats. This allowed amniotes to colonize new terrestrial environments.

The membranes of amniotic eggs also became adapted to form tough leathery or hard shells. Shells improved amniotic eggs' ability to hold in water and also made the eggs more durable and less vulnerable to small predators. Mammals, birds, reptiles, and dinosaurs are all amniotes. Although most modern mammals do not lay eggs, developing mammalian embryos still have membranes that cover them while inside their mothers.

Although amniotic eggs are watertight, they are not airtight. If they were, the eggs would suffocate. As the living cells inside an egg grow and develop, they consume oxygen and produce carbon dioxide waste (just as all animal cells do). This carbon dioxide waste needs to go somewhere, and fresh oxygen needs to be constantly supplied. Even hard eggshells are covered with tiny holes that permit gasses to be exchanged between the inside of the egg and the outside world. This need to breath places a limit on how big eggs can be.

Recall the square–cube law—as any shape increases in size, its surface area increases more slowly than its volume. Although truly enormous dinosaur eggs are often depicted in cartoons and poorly-researched science fiction, the largest known dinosaur egg is only half a meter (1.64 feet) long, and most are much smaller. Eggs that are much larger than this are not possible, because the amount of oxygen that a dinosaur developing inside an egg requires is a function of its volume, while the rate at which oxygen can be exchanged is a function of the eggshell's surface area. Giant eggs would have a low ratio of surface area to volume and would die.

# GROWTH

Hatching from relatively small eggs meant that baby dinosaurs had a lot of growing up to do. Bone histology has helped paleontologists better understand dinosaur growth rates. Recall that bone cells are called osteons. As animals grow their bones, they add osteons to their bones' outer walls. But the rate at which osteons are added is not always the same and varies with changes in growth rates. During seasonal periods, when resources needed for growth are scarce, such as during winter or the dry season, growth may slow down. This creates rings inside the bones, analogous to those of a tree trunk. These rings are called **lines of arrested growth**, or **LAGs** for short.

By studying LAGs in young and old dinosaurs, we can determine how long it took a dinosaur to grow to a particular size and at what speed a dinosaur grew. It turns out that dinosaurs grew fast. It is estimated that a *Tyrannosaurus rex* grew to its adult size in only 20 years. Even large sauropods only took 30 years to fully mature, and they are estimated to have gained an average of 1 to 2 pounds every day!

The bones of younger dinosaurs are characterized by having **high vascularity** (many blood vessels) and a texture we call **lamellar bone**. LAGs formed later, as dinosaurs grew. More mature dinosaur bone underwent a regular process called **remodeling**, where the old bone cells were replaced by newer bone cells. Bone that has been remodeled is called **haversian**, or **secondary bone**. Finally, as growth slowed and then finally stopped, a closely spaced series of LAGs formed, which is called the **external fundamental system (EFS)**. The presence of an EFS indicates that a dinosaur is skeletally mature and has stopped growing.

# ONTOGENETIC CHANGES

Some newborn or newly hatched animals look like tiny versions of their parents, and others look different in a few particular ways. For instance, newborn human babies have heads that are large relative to the overall size of their bodies and eyes that are large relative to the overall size of their heads. As humans grow up, the relative proportions of their bodies, heads, and eyes gradually approach those of their parents. Changes in the form of an organism that occur as it matures are called **ontogenetic changes**.

Big heads and big eyes are common traits of young animals. Baby dinosaurs also had relatively large heads and eyes. Some dinosaur ontogenetic changes were more dramatic. For instance, the horns and head shield of a baby *Triceratops* were short and stubby. As a *Triceratops* grew, its horns became elongated, and its frill became more expansive. Likewise, the crests of many hadrosaurs were not present in very young individuals but grew gradually as the dinosaurs reached maturity. Some ontogenetic changes involve the growth of entirely new structures. It seems that many baby ankylosaurs hatched with little or no armor and with no tail clubs. Ankylosaur body armor and tail clubs did not grow until later in life.

Changes in the relative proportions of an animal as it grows, that are not simply changes resulting from a general increase in size, are called **non-isometric** ontogenetic changes. The changes in the relative lengths of the horns and frills of ceratopsians are examples of non-isometric changes. Another example can be seen in the legs of dinosaurs like tyrannosaurs, where the tibia was much longer than the femur in juveniles, while in adults the tibia and femur were close to the same length.

**Isometric** ontogenetic changes are changes in absolute size but not proportions. For instance, unlike in tyrannosaurs, the length of ceratopsian hind legs changed proportionally as the animals grew. That is, the length of the tibia relative to the length of the femur of a baby *Triceratops* was nearly the same as the length of the tibia relative to the length of the femur of a full-grown adult.

## SEXUAL DIMORPHISM

Naturally, males and females of the same species are different—this is called **sexual dimorphism**. Sexually dimorphic features of the skeleton are usually subtle but can be extreme. Consider the massive antlers of a bull moose, which are entirely absent on females. As with the antlers of moose, it is common for sexually dimorphic features to be ontogenetic changes (after all, an animal usually does not reproduce immediately after it is born or hatches), and large, elaborate sexually dimorphic features commonly develop in males.

Sexual dimorphism is difficult to identify in dinosaurs. The ceratopsian *Protoceratops* may be sexually dimorphic, with one gender possessing larger neck shields. One sex of the ancient bird *Confuciusornis* possessed extra-long tail feathers. In both of these instances, although it is possible to distinguish two sexes, it is impossible to identify with certainty which sex corresponds to which set of dimorphic features.

Determining the sex of a dinosaur is difficult, but a recent new approach seems to have solved the problem, at least for some specimens. Laying eggs with hard shells requires a mother to donate a large quantity of calcium. In preparation for this donation, female birds grow medullary bone. **Medullary bone** contains concentrations of calcium that are stored prior to eggshell development. Bone histology work can identify medullary bone, and, because only female birds produce eggs, the presence of medullary bone shows that a particular specimen is a female. The application of this technique is limited, because medullary bone is only grown by females prior to egg production and is not present at other times.

# PARENTAL CARE

Were dinosaurs devoted parents that spent large amounts of time and energy caring for their young, or did they simply lay their eggs and leave their offspring to fend for themselves? Skeletons have been found of oviraptorosaurs (a kind of herbivorous theropod) sitting over their egg-filled nests. It appears that these dinosaurs were fossilized in the process of incubating their eggs, and it seems likely that they were also guarding their nests. Often, the skeletons of young dinosaurs are found alongside the skeletons of adult dinosaurs, and this suggests that these dinosaurs lived together as a family group. So, many dinosaurs do appear to have devoted considerable time and effort to parental care.

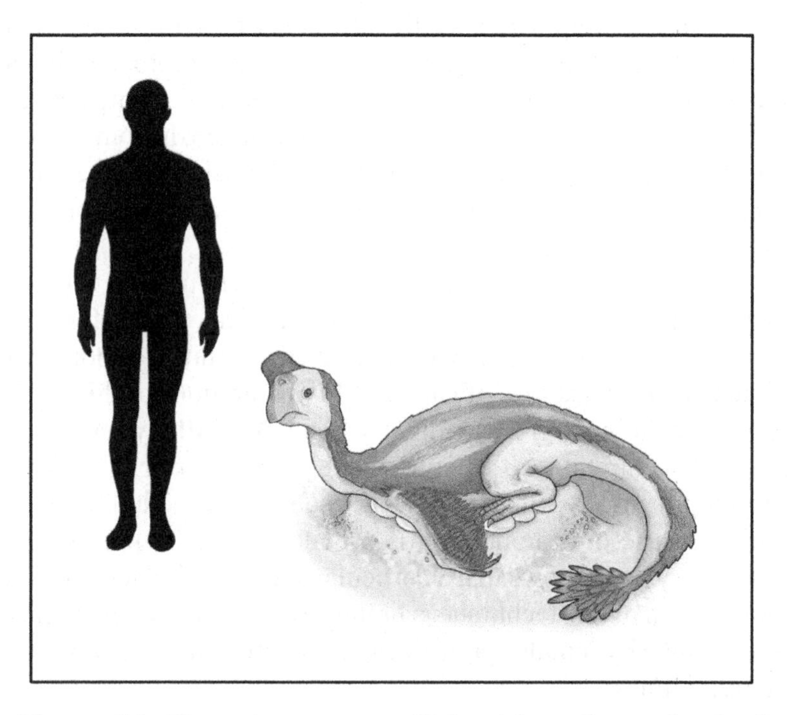

**Figure 5.1.** The oviraptorosaur *Citipati*, brooding a clutch of eggs. (Figure by Rachelle Bugeaud)

However, other lines of evidence indicate that some dinosaurs had adaptations that allowed them to avoid parental care completely. Return to our thinking about the square–cube law and the limitation that it imposes on the potential size of dinosaur eggs. Now, recall that sauropods include the largest of all dinosaurs. Some giant sauropods weighed more than ten adult elephants but laid eggs no bigger than a basketball. Although mother sauropods could not lay big eggs, they were able to lay a great many eggs. Fossil nests of sauropods from Argentina show that herds of sauropods also laid their eggs all at the same time and at the same place.

It seems likely that sauropods were using a strategy called **predator satiation**. To produce a new generation of sauropods, only a tiny fraction of the eggs that were laid needed to hatch and grow into adults. Rather than investing time into guarding and rearing their young, these sauropods may simply have produced so many offspring at one time that predators would not have been able to eat them all before they matured. This same strategy is used by many modern sea turtles.

Understanding dinosaur parental care is hard, because fossil evidence usually provides few clues about an animal's family values. As the closest living relatives of dinosaurs, modern birds and crocodilians may offer some insights. Most birds, although there are exceptions, not only care for their eggs but also feed and protect their young after they hatch. Crocodilians also tend to be good parents. Female crocodiles guard their nests and, although they do not provide their hatchlings with food, also protect their young for an extended period of time after they hatch.

CHAPTER 6

# Attack and Defense

**LEARNING OBJECTIVE FOR CHAPTER 6:** Understand defensive and offensive behaviors and structures in dinosaurs.

- **Learning Objective 6.1:** Understand examples of defensive adaptations in extant animals and potential defensive adaptations in dinosaur skeletons.
- **Learning Objective 6.2:** Understand examples of predatory adaptations in extant animals and potential predatory adaptations in dinosaurs.
- **Learning Objective 6.3:** Understand examples of intraspecific interactions in extant animals and potential intraspecific interactions in dinosaurs.
- **Learning Objective 6.4:** Understand the basic assumptions and objectives of a finite element analysis.

## PREY DEFENSES

Although dinosaurs were certainly not the insatiable bloodthirsty monsters that are commonly depicted by Hollywood, there is no doubt that dinosaurs lived with the threat or with the self-sustaining necessity of violence. Herbivorous dinosaurs had to avoid being caught and killed, and carnivorous dinosaurs had to catch and kill prey. The killing tools of carnivorous dinosaurs, like the bone-crushing jaws of tyrannosaurs and the sickle-claws of deinonychosaurs, are among the most impressive in the whole armory of the animal kingdom. Such adaptations in carnivorous dinosaurs were countered by an array of defensive adaptations in herbivorous dinosaurs.

As this book previously described (in Chapter 1), many ceratopsians evolved long horns, and ankylosaurs and stegosaurs evolved an array of spikes and armor. Having horns or spikes is a common defense strategy used by modern animals. Horns of buffalo and rhinos or the spiky quills of porcupines make them dangerous prey to attack. Even if predators succeed at killing such prey, they may be seriously injured in the process. Weapons like horns and spikes, beyond their usefulness in defending prey when attacked, are also **deterrents**. They discourage predators from choosing to attack in the first place.

Large size can be a defense entirely on its own. Giant sauropods may have lacked horns and armor, but their sheer size would have made them formidable prey. Like modern elephants, giant sauropods could have trampled even their largest potential predators, and, although not armed with spikes and clubs, many sauropods could have dealt severe blows with their massive tails.

Cursorial limbs are another obvious prey defense. Being able to outrun and/or outmaneuver potential predators keeps prey safe and avoids a physical fight altogether. Based on their hindlimb proportions, ornithomimids and many small ornithopods are cursorily adapted, and it is likely that these dinosaurs made use of their speed when threatened.

**Figure 6.1.** *Argentinosaurus*, a sauropod, may have relied on its large size to deter predators. (Figure by Joy Ang)

**Figure 6.2.** The horns of *Triceratops* were potentially lethal spears and could also be used in threat displays. (Figure by Joy Ang)

**Figure 6.3.** Covered in osteoderms and with a tail modified into a club, *Anodontosaurus*, an ankylosaur, was a walking fortress. (Figure by Joy Ang)

**Figure 6.4.** Like those of an ostrich, the long legs of *Ornithomimus*, an ornithomimid, would have given it the speed to outrun predators. (Figure by Joy Ang)

Cryptic adaptations allow potential prey to go a step further and avoid even being seen by predators. **Crypsis** is the ability of an animal to avoid detection, and cryptic adaptations include camouflage color patterns, hiding behaviors, and odor-masking chemicals. Crypsis is difficult to judge from only fossil evidence. Because cryptic adaptations are widespread among modern animals, it is reasonable to assume that cryptic adaptations were also widespread among dinosaurs. However, relying primarily on crypsis as a predator defense is more common among small animals, which are able to hide more easily behind environmental structures, than among large animals, and it is unlikely that crypsis was the sole predator defense of any large herbivorous dinosaurs.

Like crypsis, many defenses, including chemical weapons and intimidating displays, are difficult to detect from fossil evidence. Some modern animals use bright colors or false eyes to scare predators away or to clearly label themselves as toxic or otherwise dangerous. Given the diversity of dinosaurs, it is more likely that some dinosaurs used such defenses than that no dinosaurs did.

## FINITE ELEMENT ANALYSIS

Some features of dinosaurs, like horns, spikes, and teeth, are similar to structures observed in modern animals and have a form that makes their function appear obvious. However, some features of dinosaurs are novel, without comparable modern analogs, and sometimes appearances may be deceiving.

**Finite element analysis** is a technique that has been used by paleontologists to help evaluate hypotheses about the functions of many dinosaur adaptations. Finite element analyses are computer simulations that apply set material properties to a digital object and that report data on how stresses are dispersed through the object when a force is applied at a particular point.

A recent finite element analysis carried out by University of Alberta researchers attempted to evaluate the hypothesis that the tail clubs of ankylosaurs were used as weapons. The reasoning behind this study was that, if ankylosaur tail clubs were used as weapons, it is likely that the tail clubs were able to withstand large impact forces without breaking, whereas, if ankylosaur tail clubs were not used as weapons, it would be very unlikely that the tail clubs were adapted in such a way as to withstand large impact forces. The tail club of the ankylosaur *Euoplocephalus* was digitally scanned, and this digital model was then imported into a finite element analysis program.

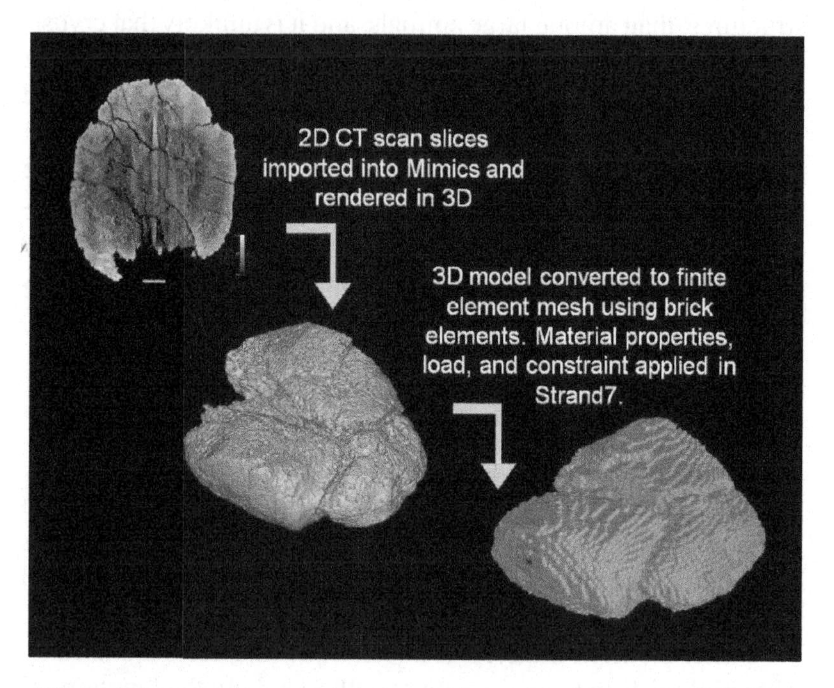


**Figure 6.5.** CT scans of ankylosaur tail clubs were converted into 3D models and tested using finite element analysis. (Figure by Victoria Arbour)

The digital tail club model was given material properties equivalent to that of bone. To simulate a tail club strike, the force of a *Euoplocephalus* tail swing was estimated and was applied to a point on the outer surface of the digital tail club model. The results showed that the resulting stresses across the tail club were insufficient to damage the club. Thus, the study concluded that ankylosaur tail clubs were capable of serving as weapons, and this supports the hypothesis that tail clubs functioned in predator defense.

## SENSES

Determining how well a dinosaur could see, hear, and smell is difficult but not impossible. The braincases of dinosaurs offer some clues, because the sizes of different regions of the brain relate directly to the strength of specific senses. Eye size can be estimated from the size of the orbits, and the inner ear cavity of the skull offers many clues to the strength of a dinosaur's hearing.

Naturally, both predators and prey benefit from keen senses that can alert the one to the presence of the other. However, the sensory needs of predators and prey are not identical. For example, herbivores, who are concerned with avoiding being snuck up on by predators, benefit from a wide field of view. For this reason, herbivores often have eyes positioned on the sides of their heads. This prevents the field of vision of one eye from redundantly overlapping with the field of vision of the other eye and maximizes how much of its surrounding an animal can see at one time.

Conversely, predators benefit from being able to maximize their perception of a single target. Often, but by no means always, predators have eyes that are positioned near each other and that both face forward. This causes the field of vision of both eyes to overlap and grants the predator **stereoscopic vision**. Stereoscopic vision allows an animal to see the same object with both eyes, and thus to see it from two slightly different angles, which improves the animal's ability to judge depth. Other animals besides predators may benefit from enhanced depth perception, and stereoscopic vision is also common among animals that fly and climb.

# PACKS AND HERDS

There can be strength in numbers, and both predators and prey may benefit from forming groups. Predators that form packs may be able to cooperatively bring down prey that is too dangerous or difficult to be attacked by an individual. It is usually more difficult to sneak up on an alert group than on a single alert individual. Prey that band together in a herd benefit from the additional sets of watchful eyes (and alert ears and noses). Some prey herds may also mount collective offenses against predators that would be too dangerous to challenge alone.

Evidence of dinosaurs forming groups comes from a variety of social display adaptations, from adaptations relating to agonistic behaviors (which will be discussed later), from trackways, and from monospecific bone beds. Some sauropod and ornithopod trackways show many sets of footprints, all from the same species of dinosaur and all heading in the same direction. These trackways suggest that the dinosaurs that made them were traveling together as a group. **Monospecific bone beds** are large accumulations of fossil bones that are all from multiple individuals of the same species. Monospecific bone beds are known to exist for many kinds of dinosaurs, including ceratopsians, hadrosaurs, and tyrannosaurs.

A healthy dinosaur ecosystem was filled with many different kinds of dinosaur species, and it is improbable that a large random fossil-sample of any such ecosystem would yield multiple bones from only a single species. Monospecific bone beds, therefore, are often interpreted as nonrandom samples, and an explanation is needed for why only one species is included. One obvious explanation for this nonrandom sample is that the particular species was traveling in a group and that the group collectively met the fate that ultimately resulted in their fossilization (such as dying in a flash flood, a mud or rock slide, or a massive volcanic ash fall).

## CHANGING STRATEGIES

Attack and defense adaptations may change ontogenetically. As an animal grows, it may become too large to effectively rely on crypsis to keep it safe, or it may become so powerful that it is able to hunt a new kind of prey. *Pinacosaurus*, a kind of ankylosaur, lacked protective armor as a juvenile, but adults grew large and heavy coverings of osteoderms. The discovery of many juvenile *Pinacosaurus* skeletons together and the discovery of many unaccompanied adult *Pinacosaurus* skeletons suggests that these ankylosaurs traveled in herds when young but were more solitary as adults.

Similarly, young tyrannosaurs have more cursorial limb proportions than do adult tyrannosaurs. This may indicate that young tyrannosaurs were adapted to hunt smaller and faster prey, while adult tyrannosaurs were better able to assault larger and slower prey. Alternatively, the more cursorial limbs of young tyrannosaurs may have been a defensive strategy that helped the youngsters avoid themselves becoming prey for adults.

## AGONISTIC BEHAVIOR

Violence is not always limited to interactions between predators and prey or even between members of two different competing species. It is common for members of the same species to fight over territory, mates, food resources, and social rank within a group. Fighting and aggressive displays between members of the same species are called **agonistic behaviors**.

Because agonistic behaviors are common, so are adaptations that facilitate them. The antlers of an elk are one example. Male elk use their antlers in head to head shoving competitions. This kind of competition that determines which of two individuals is the strongest without either combatant risking serious injury is a special kind of agonistic behavior called **ritualized agonistic combat**.

It has long been suspected that the thick domed skulls of many pachycephalosaurs were adaptations for agonistic head-butting competitions. Finite element analyses of pachycephalosaur skulls have supported this hypothesis. Like those of modern animals that engage in ritualized agonistic head-butting, such as musk oxen and bighorn sheep, pachycephalosaur skulls were strong enough to withstand severe impact forces and had special mechanical stress-reducing adaptations.

**Figure 6.6.** The pachycephalosaur *Stegoceras*. (Figure by Joy Ang)

Adaptations that serve in predator defense may also be used in agonistic behaviors. Healed puncture wounds on *Triceratops* skulls suggest that *Triceratops*, and probably many other ceratopsians, locked horns in intraspecific competitions that were likely not dissimilar to those of modern deer, cattle, and rhinos. Tyrannosaur skulls also show signs of healed wounds from agonistic behavior. **Nonlethal face biting** is a common agonistic behavior among modern carnivores, and healed bite marks on the skulls of tyrannosaurs match the tooth arrangement and form of members of their own species.

CHAPTER 7

# What Is a Species?

**LEARNING OBJECTIVE FOR CHAPTER 7:** Understand the different ways of defining what a species is and compare the strengths and weaknesses of different species concepts.

- **Learning Objective 7.1:** Understand the basic format and structure of a species name.
- **Learning Objective 7.2:** Understand the basic requirements for erecting a new species.
- **Learning Objective 7.3:** Recognize potential sources of morphological variation within a species.
- **Learning Objective 7.4:** Understand the types of evidence required to delimit species based on the biological species concept and on the morphological species concept.
- **Learning Objective 7.5:** Understand the differences between a population and a species.
- **Learning Objective 7.6:** Understand what types of evidence are available in the fossil record for defining a species.

# BINOMIAL NAMES

In the late eighteenth century, Swedish naturalist Carl Linnaeus did a great and enduring service to all biologists by introducing a new system for scientifically naming organisms. Linnaeus is considered to be the founding father of modern taxonomy. **Taxonomy** is the science of naming and organizing organisms into related groups. Prior to Linnaeus, there was not an agreed-upon system for assigning names to organisms, and this had led to considerable confusion. Under Linnaeus's system, which we still use today, every unique species of organism is given a binomial name.

**Figure 7.1.** Carl Linnaeus. (Figure in the public domain)

The **binomial name** of a species consists of two parts: the **genus name**, or generic name, and the **specific epithet**. Here are two examples, involving the dinosaur *Tyrannosaurus rex* and our own binomial name, *Homo sapiens*. *Tyrannosaurus* and *Homo* are the genus names, and *rex* and *sapiens* are the specific epithets. Note that the genus name is always capitalized and that the specific epithet is not. Also note that a binomial name is always italicized.

Organisms that are different species but that belong to the same genus (meaning that they are very similar in overall form and are more closely related to each other than to members of any other genus) have the same genus name. For instance, our close relatives *Homo erectus* and *Homo habilis* share our genus name. Specific epithets may be shared by many organisms, regardless of how closely related they are. For instance, *Tyrannosaurus rex* shares its specific epithet with *Othnielia rex* (an ornithopod dinosaur), *Nuralagus rex* (a giant extinct rabbit), *Comitas rex* (a sea snail), and *Cattleya rex* (a flower). However, the specific combination of genus name and specific epithet are not permitted to be shared by any two species.

## NEW SPECIES

There are a few other rules governing how a species gets its binomial name. The **rule of priority** states that, once a species has officially been given a binomial name, the name cannot be changed (unless it turns out that the organism is not really a new species, in which case the binomial name is abandoned). To officially give a new species a binomial name, a biologist must publish a description of the species in a widely-distributed and peer-reviewed scientific publication and must designate a holotype specimen. The published description must include a list of characteristics or combination of characteristics that makes the new species unique.

A **peer-reviewed** scientific publication is one that is not published until it has been reviewed by other scientists to verify that the contents of the publication are legitimate and scientifically reasonable. A **holotype** specimen is a physical example of the new species, and it must be kept in a research institution, such as a university or a museum, so that other scientists may study it and be able to both verify that it is a distinct species and compare it to other potentially new species that are later discovered. A holotype specimen does not necessarily need to be a complete specimen—a broken or partial specimen will do, as long as it shows the unique characters that make it a distinct species. Holotype specimens of dinosaur species are hardly ever complete.

As an example, the University of Alberta Laboratory for Vertebrate Paleontology houses the holotype specimen (UALVP 48778) of a small dinosaur called *Hesperonychus elizabethae*. The genus name is *Hesperonychus*, and the specific epithet is *elizabethae*. UALVP 48778 includes only a partial pelvis, but this pelvis provided enough information for paleontologists to determine that it represented a new kind of deinonychosaur theropod. The description of the new genus and species was published in the peer-reviewed journal *Proceedings of the National Academy of Sciences*.

**Figure 7.2.** Life restoration of *Hesperonychus*. (Figure by Joy Ang)

# SPECIES CONCEPTS

Now that we have described how a species gets its name, we come to a serious question: what exactly is a species? There is no single definition or agreed-upon concept of what a species is. The most common species concept is the **biological species concept**, which defines a species as a group of organisms that can successfully interbreed. This species concept works well when applied to most modern animals and many plants. However, it cannot be applied to the majority of modern organisms that reproduce asexually and which, therefore, cannot be said to interbreed at all. Nor can the biological species concept be applied to extinct organisms of any kind, since testing whether or not two fossils can mate is impossible.

Instead, paleontologists rely on the morphological species concept. The **morphological species concept** defines a species as a group of organisms that share a certain degree of physical similarity. In dinosaur paleontology, the morphological species concept is often applied as it relates to the biological species concept. That is, fossil specimens are assumed to belong to the same species if their physical similarities are consistent with the similarities that would be expected (based on the general pattern of physical similarity observed in modern species) between members of a group that can successfully interbreed.

A single species can sometimes be divided into separate groups by geographic barriers. Each of these geographically separate groups is called a population. A **population** is any grouping of organisms that live in the same geographic area and interbreed. One or more populations make up a single species.

# INTRASPECIFIC VARIATION

Individuals that differ in morphology because they belong to different species represent **interspecific variation**. You might think that identifying a new fossil species is as simple as pointing out the differences in the anatomy of two animals. However, there are several factors that may cause one individual to look different from another individual, even if both belong to the same species. Individuals that differ in morphology but belong to the same species are said to exhibit **intraspecific variation**.

Obviously, defining species using the morphological species concept is not an exact science, and trying to do so may be confounded by a variety of factors. Consider, for example, the taxonomic conundrum that a future paleontologist might face when discovering the skulls of an adult female moose, an adult male moose, and a juvenile male moose. Assuming that the futuristic scientist was not familiar with the sexual dimorphism and ontogenetic patterns of moose and their relatives, this paleontologist might be inclined to consider the more robust and antlered skull of the adult male moose to belong to a different species than the female and juvenile. The paleontologist might also be inclined to reason that the juvenile was just as likely to be a separate species of small deer or to be a juvenile form of either adult. Sexual dimorphism and ontogenetic change are both factors that make the morphological species concept tricky to apply, especially given the incompleteness of the fossil record.

Even among individuals of the same species, same age, and same sex, morphology varies. Take humans, for example. We come in a variety of sizes as adults, and things like eye and hair color vary among individuals. Such variation is termed **individual variation**.

An additional source of variation that needs to be considered when thinking about fossils is not biological in origin, but geological. Taphonomic processes, like plastic deformation, can change the shape of a bone, resulting in **taphonomic variation**.

# LUMPERS AND SPLITTERS

Because of these and other confounding factors, there is considerable disagreement among paleontologists over how much of a morphological difference is needed to reasonably consider one species of dinosaur as distinct from another. Paleontologists who require more differences before they consider two species to be distinct are called **lumpers**. Paleontologists who require fewer differences before they consider two species to be distinct are called **splitters**. Whether you are a lumper or a splitter can have a big effect on the total number of dinosaur species you recognize and on your interpretations of dinosaur species diversity.

Let's finish off this chapter with another look at that *Hesperonychus* specimen. *Hesperonychus* is one of the smallest known dinosaurs from North America. How did paleontologists know that it was a new species, rather than an individual of an already named species? Since it is small, could it be a juvenile of another species?

The specimen has several unique features on the pelvis that were not seen in any other deinonychosaur theropod and suggested that it was a new species. Additionally, the bones of the pelvis are tightly fused together. In juveniles, the bones of the skull, vertebrae, and pelvis are not tightly fused together, and you can see the sutures between individual bones. The sutures are not visible in the *Hesperonychus* pelvis, suggesting that it was a fully-grown individual that had a small adult size. Sexual dimorphism is harder to test, but none of the differences in the pelvis seemed likely to relate to gender-specific functions. Finally, the pelvis is well preserved, and taphonomic deformation could not have produced the unique features. So, splitting seemed to be the right choice.

CHAPTER 8

# Evolution

**LEARNING OBJECTIVE FOR CHAPTER 8:** Understand the basic theory of evolution and understand how to interpret a phylogenetic tree.

- **Learning Objective 8.1:** Understand the principles underlying the theory of natural selection.
- **Learning Objective 8.2:** Understand how shared derived characters can be used to establish relationships between dinosaur groups.
- **Learning Objective 8.3:** Distinguish between similarities resulting from convergent evolution and from shared derived characters.
- **Learning Objective 8.4:** Use a family tree to show the relationships between groups of dinosaurs.
- **Learning Objective 8.5:** Evaluate the evidence for a dinosaurian origin of birds.

# THE ORIGIN OF SPECIES

Evolution is the great unifying theory of all modern biology. The theory of evolution was first conceived by British naturalist Charles Darwin. Darwin's theory explains how new species come into existence, how organisms become adapted to their environments, and why specific groups of organisms share specific traits. It also correctly postulated that all life on Earth is related and shares a single common origin. Despite its importance and tremendous explanatory power, evolution is a simple concept and is based on four basic principles of life.

**Figure 8.1.** Charles Darwin. (Figure in the public domain)

First, many of the traits of an organism (everything from how it looks to how it behaves) are heritable. **Heritable** means that the trait is part of an organism's genetic code and, therefore, either will be, or (depending on the type of reproduction) has a chance to be, copied to the organism's offspring. Heritability is the reason that sons and daughters tend to resemble their parents. A trait must be heritable in order for that trait to evolve.

Second, sometimes organisms have heritable traits that are new, not copied from the organism's parent(s). One source for new heritable traits is random genetic mutation. For selection to occur on any given trait, there must be **variation** in that trait in a population.

Third, an organism's traits affect how successfully that organism is able to reproduce. Often, a trait's effect is indirect—that is, it improves or hinders an organism's ability to survive, which, in turn, affects how many reproductive opportunities the organism has. And fourth, natural environments have limited resources, and **competition** for these resources permits only some organisms to successfully reproduce before they die.

The theory of evolution combines these four basic principles of life together: the **differential success** of certain **variations** of a **heritable** trait, because of **competition** for limited resources, leads to the change over time of that trait in a population. If an organism is born with a new trait *(say, for example, a woodpecker has a genetic mutation that makes the keratin covering of its beak a little stronger)* and this new trait improves the organism's ability to successfully survive and reproduce *(the harder-beaked woodpecker is better able to obtain nutritious food, which improves the quality of the woodpecker's mating plumage, allows it to lay healthier eggs, and makes the woodpecker better able to feed its young)*, then that organism is more likely to outcompete other similar organisms *(the harder-beaked woodpecker consumes more food and attracts more mates while other woodpeckers starve and are mateless)*.

The beneficial trait will likely be inherited by the organism's more abundant offspring *(many clutches of harder-beaked woodpeckers hatch)*. Over time and generations, the beneficial trait is likely to become widespread *(harder-beaked woodpeckers thrive, while softer-beaked woodpeckers may eventually become extinct)*. At the same time that one beneficial trait is becoming widespread, so, too, may many other beneficial traits become the norm *(for example, sharper woodpecker claws that can better hold onto tree bark, more sensitive woodpecker ears that can better detect boring insects, broader wings that can better maneuver through forests, or an additional mutation that helps to make the beaks even harder)*. Eventually, when many new traits become widespread, the organisms that have these accumulated new traits *(the super-hard-beaked, sharp-clawed, sensitive-eared, broad-winged woodpeckers)* are so different from their ancestors *(the soft-beaked, dull-clawed, insensitive-eared, narrow-winged woodpeckers)* that they constitute a new species.

The evolution of a new species does not necessarily require the extinction of its ancestor. For instance, a new species might simply branch off from an ancestral species if only a single population of the ancestral species was exposed to new environmental conditions that favored new traits. While the population in the new environment would acquire new traits better adapted to that environment and evolve into a different species, the ancestral species might continue to exist in its ancestral environment.

Thus, evolution uses only basic principles of the natural world to explain how one species can give rise to others. It is often mistakenly said that the evolution of new species is a random process. Just the opposite is true. While new traits may be introduced by random mutations, the determination of which traits are successfully passed on to later generations is not random. Instead, it is based on a specific criterion: how well each trait improves an organism's reproductive success. Of course, most random mutations are more likely to diminish than they are to improve an organism's success. The competitive selection process by which detrimental traits are competitively discarded and advantageous traits are retained is called **natural selection**.

# CHARACTERS

Evolution provides a framework that modern taxonomy uses to categorize organisms. Organisms are grouped together based on their most recent **shared common ancestors**. For instance, all hadrosaurs, ceratopsians, ankylosaurs, and stegosaurs share a more recent common ancestor with each other than they do with sauropods, prosauropods, and theropods. Thus, hadrosaurs, ceratopsians, ankylosaurs, and stegosaurs are grouped together, and we call this group the Ornithischia. All ornithischians, sauropodomorphs, and theropods share a more recent ancestor with each other than they do with all other amniotes, and we call this group Dinosauria.

In other words, all dinosaurs are classified together in a group because all dinosaurs evolved from a single species of amniote. Within the dinosaurs, all ornithischians are classified together because they evolved from a single particular species of dinosaur, while theropods are classified together in a different group because they evolved from another particular species of dinosaur.

How can we figure out which dinosaurs share a more recent common ancestor, and, therefore, should be grouped together? This is done by studying characters. A **character** is any heritable trait that can be described and labeled. A **shared derived character** is a character that is present in two or more groups and their common ancestor but is not present in any more distantly related groups. A shared derived character is also called a **synapomorphy**. For instance, all species of ornithischians have a special bone in the lower jaw that forms a beak, called the predentary, and no other dinosaurs have this special beak bone. Thus, the character of the predentary was passed on to all ornithischians from their ancient shared ancestor and is a synapomorphy that testifies to this shared ancestry and can be used to define the group.

## CONVERGENT EVOLUTION

Identifying shared derived characters and using them to identify evolutionarily-related groups sounds easy, but often it is not. One of the biggest challenges to determining evolutionary relationships is the common phenomenon of convergent evolution. Recall that spinosaurs are a kind of theropod dinosaur with long narrow snouts and conical teeth. Long narrow snouts and conical teeth are both characters that are considered to be synapomorphies that define spinosaurs as a special group among theropods. But crocodiles also have long narrow snouts and conical teeth. Should crocodiles then be considered as a type of spinosaur, or should we classify spinosaurs not as theropods but as crocodiles? The answer is no. Why?

Although crocodiles and spinosaurs are similar in the shapes of their jaws and teeth, these similarities are not the results of a shared history of inheritance. The evolution of similar traits in two different lineages is termed **convergent evolution**. Usually, convergent evolution results when two lineages must adapt to similar environments and to similar modes of life.

You will remember that, like crocodiles, spinosaurs are thought to have been partially piscivorous. The need to adapt to feeding on aquatic prey is probably responsible for the similarity in the tooth and jaw structures of the two groups. Although convergent evolution is troublesome for paleontologists when they try to classify dinosaurs, it is one of the greatest tools they have when it comes to trying to figure out a dinosaur's ecological role, because identifying convergent similarities between dinosaurs and modern animals allows paleontologists to infer similar lifestyles.

## PHYLOGENETIC ANALYSIS

Suppose that in the previous example you did not know that narrow jaws and conical teeth are traits typical of piscivores, and, therefore, you had no particular reason to suspect that these similarities between spinosaurs and crocodiles were the result of convergent evolution. How could you figure out that spinosaurs are theropod dinosaurs and not crocodiles?

The answer is obvious. Although spinosaurs do uniquely share some traits in common with crocodiles, they share many more traits in common with other dinosaurs (such as an erect stance, a special open hip socket, additional processes in the cervical vertebrae, hinge-like ankle joints, and numerous subtle features of the skull and hindlimb musculature) and still more with theropods in particular (such as a flexible lower jaw, a reduced number of digits in the hands and feet, thin-walled limb bones, and fused clavicles). It is simpler to assume that the two characters in common between spinosaurs and crocodiles are the result of convergence than it would be to assume that the huge number of similarities between spinosaurs and theropod dinosaurs are all the result of convergence. The idea that "all other things being equal, the simplest answer is usually the right one" is called **parsimony**. Parsimony is also referred to as **Occam's razor**.

Figuring out the relationships between only a few kinds of distantly related organisms is usually not too hard, as we can rely on our own ability to judge which of a handful of alternative relationships is the most parsimonious. To determine the evolutionary relationships between large numbers of species, many of which may be closely related, and to take into consideration a large number of characters, paleontologists use computer programs. These programs analyze a list of characters that is first compiled by the researcher. This list is called a **character matrix**. Based on the character matrix, the computer program applies the principle of parsimony to arrange the organisms in a sequence of relationships that requires the fewest instances of convergent evolution. The resulting arrangements look like diagrams of a "family tree" and are called **phylogenetic trees**.

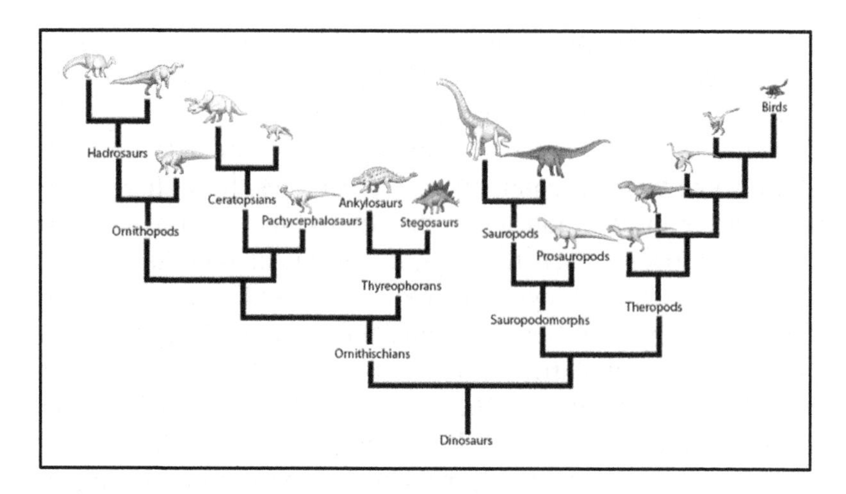


**Figure 8.2.** The dinosaur family tree. (Figure by Joy Ang, Rachelle Bugeaud, Veronica Krawcewicz, and W. Scott Persons)

Phylogenetic trees are composed of **nodes** and **branches**. A node is where two branches diverge and shows the point at which two lineages shared a common ancestor. After a node, the pattern of subsequent branches and nodes shows how the descendants of that common ancestor continued to diverge from each other. A group of species that share a common node is called a **clade**. Clades can be very small (even as small as two species), or very large—there are no size limits.

In school biology courses, you may have been taught a system of classification that we call the Linnaean Hierarchy, which classifies organisms as belonging to a Kingdom, Phylum, Class, Order, Family, Genus, and Species. While the Linnaean hierarchy works pretty well, especially considering it was formed long before our current ideas about natural selection and evolution had solidified, the original classifications do not always work well with our current understandings of the evolutionary relationships of organisms.

For example, Mammalia (mammals), Aves (birds), Amphibia

(amphibians,) Reptilia (lizards, snakes, crocodiles, and turtles), and Pisces (fish) were all classified at the equivalent rank of Class. But we can trace all four-limbed animals (the tetrapods, including mammals, reptiles, birds, and amphibians) back to a fish ancestor—Class Mammalia, Aves, and Amphibia are all a subset of Class Pisces (now called Osteichthyes). One of the best examples of this problem is what to do with the birds, as the following section discusses.

# BIRDS ARE DINOSAURS

When the theory of evolution was first proposed by Charles Darwin in the eighteen hundreds, many scientists were skeptical and wanted to see more evidence supporting evolution in the fossil record. These skeptics reasoned that, if indeed new species evolved from ancient species, and if all life shared a common evolutionary history, then fossils should exist of major "missing links"—that is, organisms that show an evolutionary connection between two major groups of organisms by displaying some traits that are characteristic of one group and some traits that are characteristic of the other group.

Thomas Henry Huxley was a close colleague of Charles Darwin and one of the earliest advocates for the theory of evolution. Huxley was also the first scientist to recognize that birds evolved from dinosaurs. And he cited the newly discovered specimens of ***Archaeopteryx*** as fossils of a "missing link" between dinosaurs and birds.

**Figure 8.3.** Thomas Henry Huxley. (Figure in the public domain)

Specimens of *Archaeopteryx* had been found exquisitely preserved in fossil lake deposits. These specimens clearly show that *Archaeopteryx* has long wing-feathers and tail-feathers just like a bird, but they also show that *Archaeopteryx* had teeth, clawed fingers, and a long series of tail vertebrae just like a dinosaur.

With the help of *Archaeopteryx*, Thomas Henry Huxley showed that transitional forms do exist in the fossil record, just as the theory of evolution predicted, and also showed that birds are a branch of the dinosaur family tree.

**Figure 8.4.** The skeleton and feather impressions of *Archaeopteryx*. (Figure by W. Scott Persons)

Of course, not everyone was convinced by Huxley's arguments. It took many years before the theory of evolution was fully accepted by the entire field of biological science and even longer for the theory of a dinosaur origin of birds to be fully accepted. Since Huxley's time, paleontologists have come to recognize an increasing number of characters that birds and other theropod dinosaurs share. One of the most significant shared characters was discovered in a specimen of the little dinosaur ***Sinosauropteryx***.

*Sinosauropteryx* was the first non-avian (non-bird) dinosaur to be discovered with feathers. The feathers of *Sinosauropteryx* had a simple structure compared to the feathers of modern birds and were used for insulation, not for flight; but they were feathers just the same. Many other small theropod specimens have since been found with feathers, some with complex flight feathers. Feathers have also been found on specimens of the large tyrannosauroid ***Yutyrannus***, showing that some large dinosaurs had them as well.

Birds are the only clade of dinosaurs alive today. Birds can be classified as theropod dinosaurs, because they evolved from theropod dinosaurs. But where we draw the line between non-avian theropod dinosaurs and "birds" is more complicated than you might think. So many fossils of bird-like dinosaurs and dinosaur-like birds have been found that the line between them is blurry. The clade name "Aves" (birds) can be interpreted in several different ways, and it is not easy to say exactly when dinosaurs started being birds.

Many paleontologists consider the term "bird" to be synonymous with the evolution of flight. For them, "birds" includes the first dinosaur to fly and all of its descendants. This definition is tricky, because we cannot always tell whether a dinosaur was capable of flight. Many dinosaurs (such as the oviraptor *Caudipteryx*) had wing-like arrangements of feathers on their arms but were clearly land-bound and presumably used their arm feathers in courtship displays. Also, flight did not evolve in one great evolutionary leap. Many feathered dinosaurs first became adapted to simple gliding. They jumped and sailed through the forests like modern flying squirrels do. It is even possible that more than one of these dinosaur lineages evolved true flight, but only one could have been the lineage that gave rise to modern birds.

Other paleontologists argue that we can define birds by deciding that *Archaeopteryx* was the first bird and that "birds" includes *Archaeopteryx* and all of its descendants. But it may not be accurate to think of *Archaeopteryx* as the ancestor of all birds. *Archaeopteryx* may be an evolutionary side branch very closely related to birds but not the dinosaur from which all birds eventually evolved. Some phylogenetic studies suggest that, although *Archaeopteryx* is very close to the evolutionary split that led to birds, it falls just on the other side, closer to deinonychosaurs.

CHAPTER 9

# Stratigraphy and Geologic Time

**LEARNING OBJECTIVE FOR CHAPTER 9:** Understand basic stratigraphic concepts, the scale of earth history, and the evolution of dinosaurs through time.

- **Learning Objective 9.1:** Understand the geological Principle of Superposition.
- **Learning Objective 9.2:** Describe important events in the geological history of Earth.
- **Learning Objective 9.3:** Classify types of dinosaurs based on the geological age in which they were most common.

## STRATIGRAPHY

As this book previously discussed (in Chapter 2), sedimentary rocks form from small organic or inorganic particles (called sediments) that accumulate and are cemented or compacted together. As sediments are deposited, they gradually build up on top of each other in layers. Over time, deep sequences of layered sedimentary rocks result. In such a sequence, the oldest rocks (which are formed from the oldest deposited sediments) are at the bottom, and the layers become increasingly younger toward the top. The tendency for rock layers to be chronologically stacked is called the **Principle of Superposition**.

**Figure 9.1.** Alternating light and dark sedimentary layers at Dinosaur Provincial Park. (Figure by W. Scott Persons)

The Principle of Superposition is only an expression of how things normally work, and there can be exceptions. Igneous rocks that form from volcanic activity may vertically cut through horizontally-arranged layers of rocks, and mountain-building events may tilt, fold, and even flip rock layers. **Stratigraphy** is the science of using the arrangement and composition of rock layers to interpret geological history. A large uninterrupted sequence of rock that is made of multiple layers that all share similar properties (such as mineral composition and average sediment grain size) and that all formed under similar conditions is termed a **formation**. When a sequence of rock changes from one formation to another, it indicates that a large-scale change occurred in the environment where the rocks were being deposited.

# RADIOMETRIC DATING

The Principle of Superposition allows a stratigrapher to infer the relative age of rock layers (that is, how old one layer is relative to another), but it does not determine the absolute age (that is, how old in years the layers are). To age rocks in absolute terms, a technique called **radiometric dating** is used.

All matter (including rocks) is composed of chemical elements, which are atoms composed of a particular number of protons and electrons (positively and negatively charged particles). Some chemical elements are also composed of neutrons (neutrally charged particles), and some of these chemical elements may exist as isotopes. An **isotope** is a variant of a chemical element that has an unusual number of neutrons. Some isotopes are unstable and will undergo radioactive decay, whereby energy is released and a new atom (or atoms) with a different composition of particles results. These resulting atoms with different particle compositions are called the **decay products**. At what time a single isotopic atom will undergo radioactive decay is unpredictable, but a large collection of isotopes will radioactively decay at a mathematically predictable rate.

When a new rock forms, it has a ratio of isotopes and decay products that matches that of the environment. As the rock ages, the isotopes decay and the ratio of isotopes to decay products decreases. Using a special machine called a mass spectrometer, it is possible to measure the isotope ratio of a rock, and this ratio can tell you how long ago the rock formed.

Unfortunately, sedimentary rocks are never really "new"—that is, they are made of sediments that had already formed and that were already potentially undergoing radioactive decay. Thus, sedimentary rock cannot usually be radiometrically dated. On the other hand, igneous rocks *are* formed anew and can usually be radiometrically dated. But if fossils are usually found in sedimentary rocks and not in igneous rocks, how can we ever tell how old a fossil and its rock layer are?

The answer is: by combining radiometric dating and the Principle of Superposition. If sedimentary rocks that contain fossils are found between two horizontally-deposited layers of igneous rocks, then dating the igneous rocks above the sedimentary layer will tell us what age the fossils must be older than, and dating the igneous rocks below the sedimentary layer will tell us what age the fossils must be younger than. So, we can confidently bracket the age of the fossils.

In some instances, it is even easier to date fossils, because we can be certain that particular particles of igneous rocks that compose the sedimentary rocks that the fossils are buried in were incorporated into the sediment at nearly the same time that they were formed. For instance, fossils may be buried by, or be buried near, deposits of volcanic ash. Volcanic ash forms at the moment of an eruption, and the time between when an eruption occurs and when its ash falls to the surface is inconsequently short. Volcanic ash deposits are a key tool in fossil dating.

## THE GEOLOGIC TIME SCALE

Stratigraphy and radiometric dating are combined to piece together the history of the earth and to create the geologic time scale. **The geologic time scale** is a standardized series of chronological divisions that parses Earth's history into discrete named units. The largest units in the time scale are Eons, followed by Eras, Periods, and Epochs. You can download the official geologic timescale from the International Commission on Stratigraphy, at the following URL: http://www.stratigraphy.org/index.php/ics-chart-timescale. What follows in the remainder of this chapter is a brief overview of the geologic time scale and of some of the important events that occurred during its different divisions.

## THE HADEAN EON—4.6 TO 4 BILLION YEARS AGO

The Hadean Eon is named for Hades (the Greek god of the underworld), and by the beginning of this eon, the rest of the universe was already over 9 billion years old. The formation and early years of the earth were a tumultuous time, with the surface of the earth partially molten and with volcanic activity widespread. At roughly 4.5 billion years ago, the young earth collided with a smaller planetoid. This collision ejected a large mass of debris, which was held in orbit by the earth's gravity and eventually formed the moon.

By the end of the Hadean, the earth had cooled, and large oceans covered much of its surface. Complex organic molecules and the earliest true life forms are thought to have formed in these early oceans. The oldest rocks on Earth have been dated at only about 4.4 billion years old, though rocks discovered on the moon are older.

## THE ARCHEAN EON—4 TO 2.5 BILLION YEARS AGO

The oldest known fossils come from the Archean Eon. These fossils are of simple single-celled organisms. More advanced forms later evolved in the Archean, including cyanobacteria. The cyanobacteria were photosynthetic and eventually produced large amounts of oxygen gas, which became concentrated in the earth's atmosphere. Some cyanobacteria formed structures called stromatolites, which are some of the best fossil records of early life. Stromatolites look like lumpy stones, but when you cut them in half, you can see the layers that were created as the cyanobacteria secreted sticky films that trapped particles of sediment.

## THE PROTEROZOIC EON—2.5 BILLION TO 541 MILLION YEARS AGO

At approximately 1.7 billion years ago, the first multicellular organisms evolved. Because single-celled and early multicellular life had no bones or other hard parts and was usually microscopic, the fossil record of this early life is poor. Within the Proterozoic, the time span from 630 to 542 million years ago is known as the Ediacaran Period. During the Ediacaran, large forms of life with some harder parts evolved, including the first animal life.

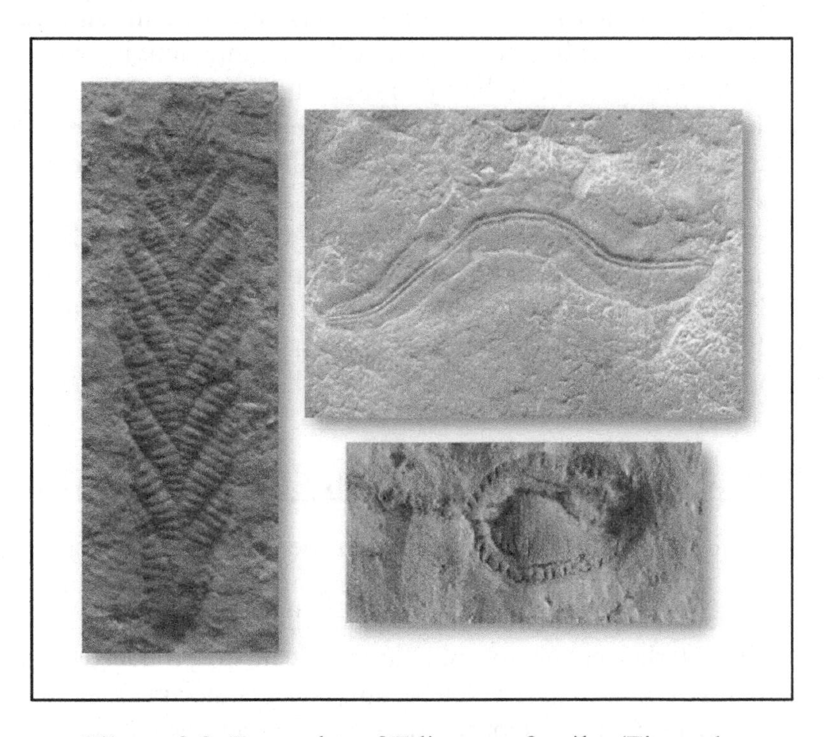

**Figure 9.2.** Examples of Ediacaran fossils. (Figure by W. Scott Persons)

## THE PHANEROZOIC EON—541 TO 0 MILLION YEARS AGO

The Phanerozoic Eon is subdivided into three eras, which are themselves subdivided into numerous periods. It is during the Phanerozoic that animal life rapidly evolved into a multitude of diverse forms, including dinosaurs. We will consider the events of each Phanerozoic era and period.

## The Paleozoic Era—541 to 252 million years ago

At the start of the Paleozoic, animal life was restricted to primitive invertebrates living in the oceans, but by its close, great forests covered the land and teamed with reptiles, amphibians, and insects.

### The Cambrian Period—541 to 485 million years ago

The beginning of the Cambrian marks such a dramatic diversification of aquatic animal life that it is often referred to as **The Cambrian Explosion**. Sponges, mollusks, worms, and many kinds of arthropods (including trilobites) evolved. By the end of the Cambrian, invertebrate land animals had evolved and so had the first vertebrate animals.

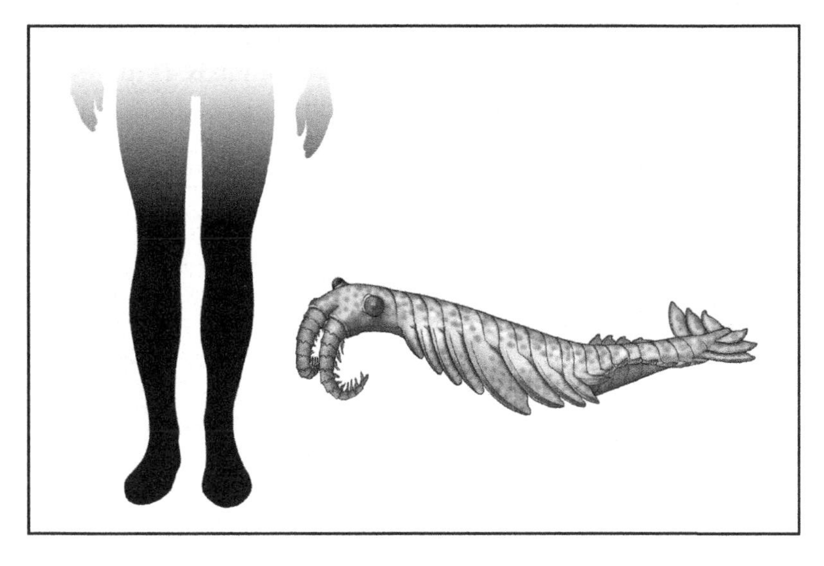

**Figure 9.3.** The invertebrate *Anomalocaris* was once the top predator in the Cambrian oceans. (Figure by Rachelle Bugeaud)

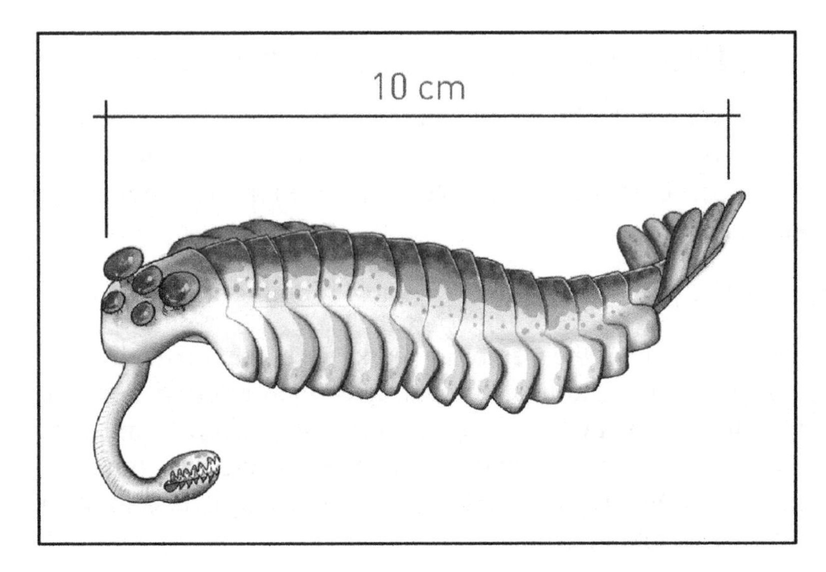


**Figure 9.4.** With a bizarreness characteristic of Cambrian life, *Opabinia* remains a mystery in terms of both its life habits and its evolutionary relationships to modern animals. (Figure by Rachelle Bugeaud)

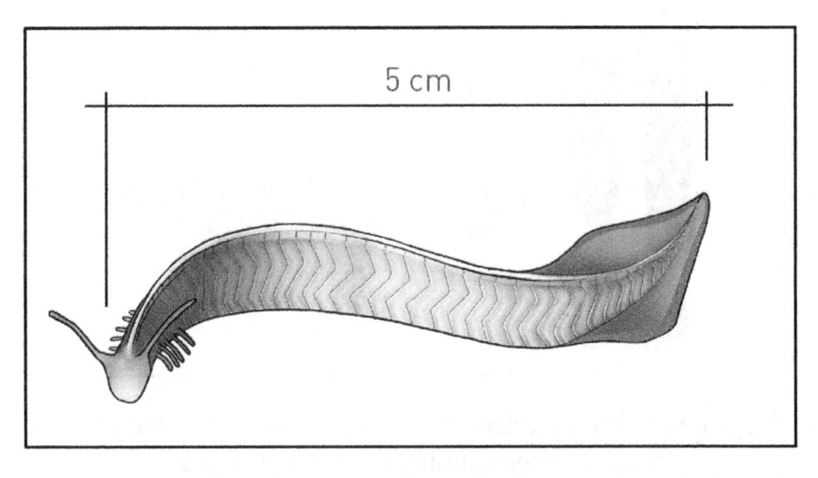


**Figure 9.5.** Though it lacked a hard skeleton, *Pikaia* is thought to have been an early ancestor of vertebrate animals. (Figure by Rachelle Bugeaud)

### The Ordovician Period—485 to 443 million years ago

Global sea levels were high. Life in the oceans continued to diversify, with fish increasingly becoming the dominant large aquatic animals.

### The Silurian Period—443 to 419 million years ago

Until this point, fish had not yet evolved jaws. With the evolution of jaws during the Silurian came the evolution of large predatory fish. Primitive plant life began to flourish on land.

### The Devonian Period—419 to 359 million years ago

The first forests appeared on land. The first land vertebrates evolved and soon gave rise to the first tetrapods.

### The Carboniferous Period—359 to 299 million years ago

Amphibians were widespread in abundant swamps, and reptiles, the first amniotes, evolved. Much of the coal that is mined today formed from the rotting plants of Carboniferous swamps.

### The Permian Period—299 to 252 million years ago

The continents collided together and formed a single supercontinent called **Pangaea**. Reptiles evolved into three main lineages: the anapsids, the synapsids (which would go on to evolve into mammals), and the diapsids (which would go on to evolve into lizards, snakes, crocodilians, and dinosaurs). Many of the terrestrial rocks from this period of time represent dry desert environments. The single greatest mass extinction in our planet's history occurred at the end of the Permian.

## The Mesozoic Era—252 to 66 million years ago

The Mesozoic is often referred to as the Age of Dinosaurs. It is during this time that dinosaurs evolved and became the dominant form of large terrestrial animal life. The Mesozoic was also a time when many kinds of marine reptiles evolved, including the ichthyosaurs, plesiosaurs, and mosasaurs. The first true turtles, crocodilians, lizards, snakes, mammals, and birds evolved at this time as well. The Mesozoic Era has been, and will continue to be, examined and discussed throughout this book.

Many dinosaur books and movies depict dinosaur species from vastly different periods existing at the same time and place. This can be a big mistake! The non-avian dinosaurs existed for over 160 million years, meaning that there is less time separating the first humans from the last dinosaurs than there is separating the last dinosaurs from the first dinosaurs, and not all clades of dinosaurs were present throughout the entire Mesozoic.

### The Triassic Period—252 to 201 million years ago

During the first 10 million years of the Triassic, life gradually recovered from the mass extinction that occurred at the end of the Permian. The first mammals and dinosaurs evolved during the later portion of the Triassic. The supercontinent, Pangaea, began to break apart. Many of the dinosaurs from this period of time look fairly similar to each other. The first dinosaurs were small, bipedal, and carnivorous. Prosauropods were some of the first large herbivorous dinosaurs.

As dinosaurs were evolving to fill large-bodied ecological roles on land, other amniote groups were evolving to fill them in the sea and air. It was in the Triassic that the first **ichthyosaurs** evolved. The name "ichthyosaur" literally means "fish lizard," but ichthyosaurs are not lizards, and they certainly aren't fish. Even so, the name "ichthyosaur" still seems fitting, because they are a group of reptiles that took on a fish-like lifestyle and evolved a very fishy body form.

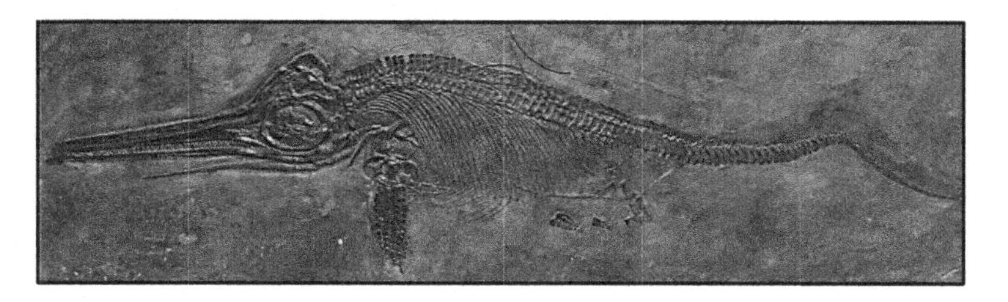

**Figure 9.6.** Ichthyosaur skeleton. (Figure by W. Scott Persons)

The ancestors of ichthyosaurs were fully terrestrial reptiles, but just like the ancestors of modern whales, dolphins, seals, and sea turtles, the group found success by making an evolutionary return to the water. To adapt to an aquatic life, ichthyosaurs evolved paddle-like front and hind limbs, a finned tail, and even a shark-like dorsal fin. The long snouts of most ichthyosaurs resemble those of dolphins and are filled with conical teeth—good equipment for a piscivorous diet. Despite their many fish-like adaptations, ichthyosaurs never evolved gills and still needed to come to the surface in order to breathe air.

Late into the Triassic, ichthyosaurs were joined in the seas by another group of reptiles that also evolved a secondarily aquatic lifestyle: the **plesiosaurs**. Most plesiosaurs had large chests and torsos, broad paddle-shaped limbs, and relatively short tails. In front of their shoulders, plesiosaurs varied tremendously. Some had short necks and huge jaws, and others had elongated serpentine necks with small heads.

The Triassic also saw the first vertebrates to conquer the air. These were not dinosaurs (that wouldn't happen until birds evolved later in the Jurassic). **Pterosaurs**, or as they are commonly called "pterodactyls" or simply "dactyls," are close relatives of dinosaurs and branched off from the reptilian family tree at roughly the same time that dinosaurs did. Unlike birds, which have arms that support wings made of feathers, and bats, which have wings made from skin stretched between multiple fingers, pterosaurs have membranous wings supported by a single extremely elongated finger.

Early pterosaurs belong to a group called **rhamphorhynchoids**, which were common in the Late Triassic and throughout the Jurassic. Rhamphorhynchoids are all small pterosaurs, with the largest forms approaching the size of a modern raven or hawk. Rhamphorhynchoids are also characterized by long tails. Some rhamphorhynchoid tails terminated in flaps of skin called tail vanes. The function of these tail vanes is unclear, but it is hypothesized that they acted as in-flight rudders or stabilizers and/or were structures used in sexual displays. (The Triassic will be discussed in more detail in Chapter 11.)

## The Jurassic Period—201 to 145 million years ago

The Jurassic was the peak of sauropod diversity, and sauropods were now the dominant terrestrial herbivores. Small and medium-sized ornithopods were common. Non-coelurosaurian theropods, like *Allosaurus*, were the dominant terrestrial carnivores. The stegosaurs were almost completely restricted to the Jurassic, and the first ankylosaurs and ceratopsians appeared at this time, although they were not particularly abundant or diverse. The first birds, including *Archaeopteryx*, evolved during the Jurassic. The Morrison Formation of the western United States has provided one of the most famous records of dinosaurs from this time.

In the Jurassic, rhamphorhynchoid pterosaurs gave rise to a new pterosaur group: the pterodactyloids. **Pterodactyloids** differed from rhamphorhynchoids in the morphology of their tails, which were short, and the carpels in their wrists, which were elongated and made a greater contribution to the length of the wing. Unlike rhamphorhynchoids, many pterodactyloids had large head crests, which were presumably display structures. There were many species of small pterodactyloids, some smaller than a robin; but some species had wingspans of over 10 meters (32.81 feet), making them the largest animals to ever fly.

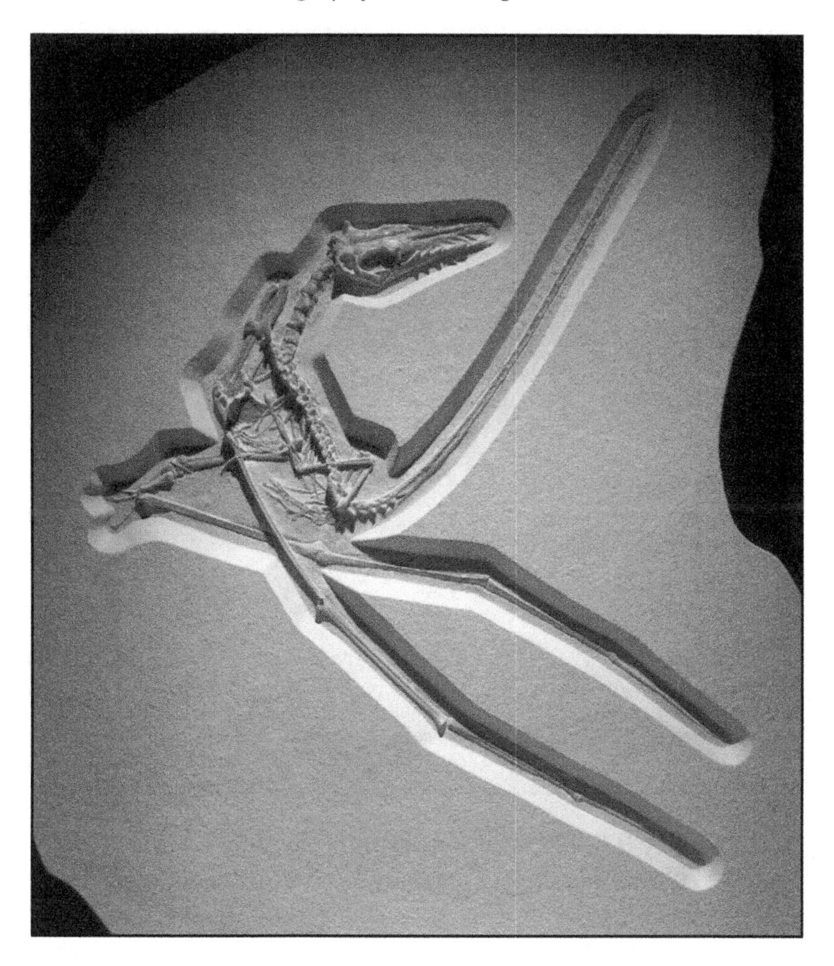

**Figure 9.7.** Skeleton of the Jurassic rhamphorhynchoid *Rhamphorhynchus*. (Figure by W. Scott Persons)

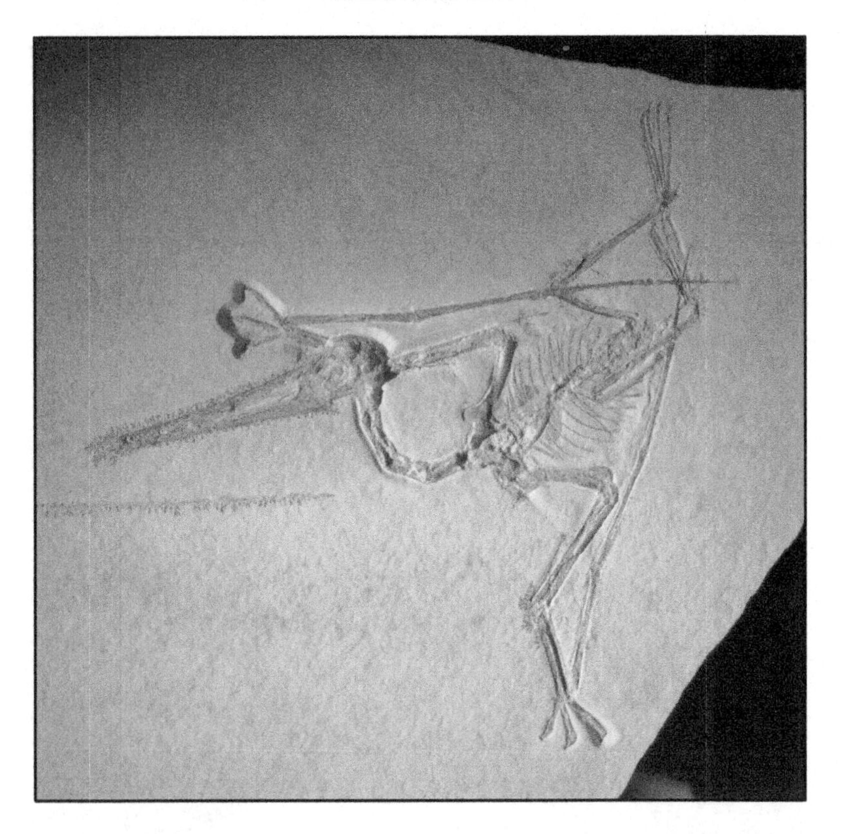

**Figure 9.8.** Skeleton of the Jurassic pterodactyloid *Pterodactylus*. (Figure by W. Scott Persons)

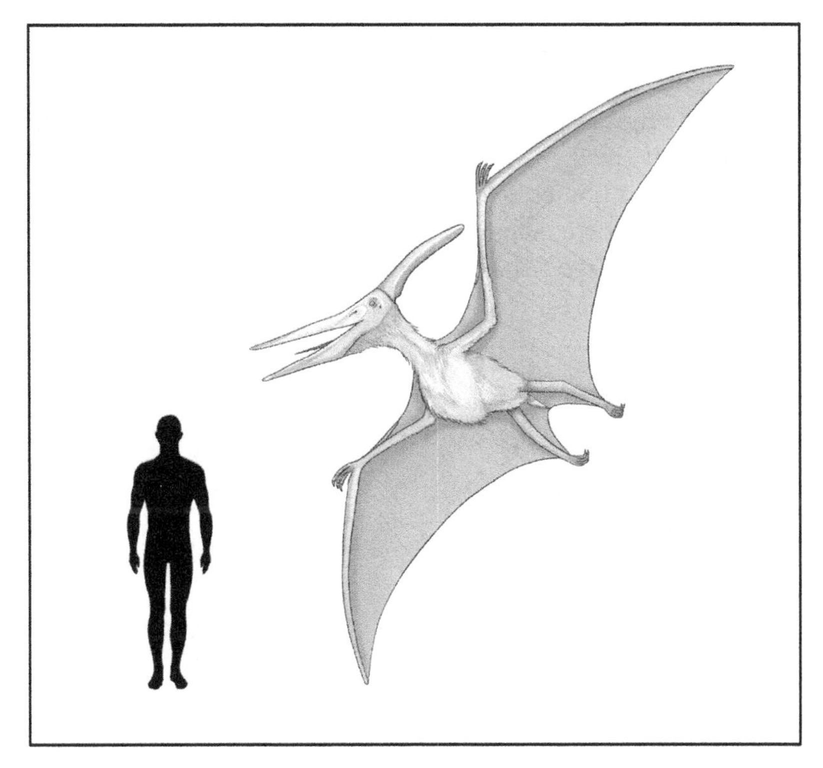

**Figure 9.9.** The large pterodactyloid *Pteranodon*. (Figure by Joy Ang)

## The Cretaceous Period—145 to 66 million years ago

In the Cretaceous, dinosaurs continued to diversify, and flowering plants became common. The Yixian Formation of China, the Wealden Supergroup of England, and the Cedar Mountain Formation of Utah are important Early Cretaceous fossil-rich rock units. Often considered the apex of non-avian dinosaur diversity, it was in the Late Cretaceous that many of the most famous dinosaurs evolved.

Sometime in the Cretaceous Period, a third major reptilian group began patrolling the Mesozoic waters. **Mosasaurs** were relatives of modern monitor lizards and snakes. Like ichthyosaurs and plesiosaurs, mosasaurs had tail fins and limbs modified into paddles, but the bodies and tails of mosasaurs were more elongate. Many mosasaurs are the right size to have preyed on small and medium-sized fish, but some were true sea monsters with huge jaws and bodies over 18 meters (59.06 feet) long. These aquatic giants seem adapted for deep-sea big game hunting, and they likely ate large fish and other marine reptiles.

**Figure 9.10.** Skull of the giant mosasaur *Tylosaurus*. (Figure by W. Scott Persons)

**Figure 9.11.** The plesiosaur *Elasmosaurus*. (Figure by Joy Ang and Veronica Krawcewicz)

At the end of the Cretaceous, a large meteor collided with Earth, and this event, along with its catastrophic consequences, is thought to have brought about a mass extinction, which killed all non-avian dinosaurs. Ichthyosaurs and rhamphorhynchoid pterosaurs became extinct before the close of the Cretaceous, but pterodactyloid pterosaurs, plesiosaurs, and mosasaurs shared the dinosaurs' fates and were all wiped out by the end-Cretaceous mass extinction event.

## The Cenozoic Era—66 to 0 million years ago

The Cenozoic is often referred to as the Age of Mammals. Although mammals had been around since the Triassic, the extinction of the dinosaurs (except for birds) allowed mammals to evolve larger forms and to fill many new ecological roles. Grasses only become abundant at this time.

### The Paleogene Period—66 to 23 million years ago

Global temperatures began to cool. Mammals diversified into a variety of new forms, including primates, bats, and whales. Birds also diversified.

### The Neogene Period—23 to 2.6 million years ago

Global temperatures continued to cool. The first hominids evolved in Africa.

### The Quaternary Period—2.6 to 0 million years ago

Earth experienced several large glaciation events, or "ice ages," the first anatomically modern humans evolved, and human civilization spread.

CHAPTER 10

# Paleogeography, Plate Tectonics, and Dinosaur Diversity

**LEARNING OBJECTIVE FOR CHAPTER 10:** Understand the basic concepts of plate tectonics and the evolution of the earth's surface.

- **Learning Objective 10.1:** Understand the basic evidence for plate tectonics and the mechanisms underlying it.
- **Learning Objective 10.2:** Understand the characteristics of the layers of the earth.
- **Learning Objective 10.3:** Recognize large-scale paleogeographic features like Pangaea, Laurasia, and Gondwana.
- **Learning Objective 10.4:** Understand how the sequence of continental drift influenced the evolution of dinosaurs.
- **Learning Objective 10.5:** Understand the geographic effects of a warmer global temperature during the Mesozoic Era.

# CONTINENTAL DRIFT

In 1912, a German researcher named Alfred Wegener drew the scientific community's attention to several curious facts. Wegener noticed that the eastern coastline of South America and the western coastline of Africa looked like two connectable puzzle pieces. He also noticed that the fossils of many ancient animals (which, as far as anyone could tell, were not animals that would have been capable of swimming across the Atlantic Ocean) could be found in both South America and Africa and that several geologic formations in South America had seemingly identical twins in Africa.

Wegener suggested that Africa, South America, and possibly other continents had once been connected and had since drifted apart. Wegener's reasoning was sound, and his evidence was tantalizing, but his theory of continental drift had a huge hole in it: Wegener could not offer a convincing mechanism for how landmasses as big and as seemingly immobile as continents could move. Many years later, Wegener's idea of moving continents was vindicated, and an explanation for how such a massive phenomenon occurs was discovered, as the following section discusses.

**Figure 10.1.** Alfred Wegener. (Figure in the public domain)

# PLATE TECTONICS

Below its surface, the earth is not a uniform mass of rock. The outermost layer of the earth consists of the continents and ocean basins and is called the **crust**. The thickness of the crust varies but is usually between 5 and 25 kilometers (3.11 and 15.53 miles) deep. By comparison to the other layers of the earth, the crust is thin.

Below the crust is a layer called the mantle. The **mantle** is a layer over 2,500 kilometers (1,553.43 miles) deep. The uppermost portion of the mantle is solid. Along with the crust, this solid upper portion of the mantle is called the **lithosphere**. The lithosphere is not one unbroken layer but is actually composed of many discrete pieces, or plates, that fit together. Below the lithosphere is a portion of the mantle called the **asthenosphere**. While the lithosphere is rigid, the asthenosphere is viscous, slowly flowing, and its shape may be deformed under the uneven weight of the lithosphere. Although it flows, the mantle is not a liquid but rather a viscous solid. The intense heat and pressure at great depths cause the solid mantle to behave like a fluid—similar to plasticine or molding clay that is a solid at rest but squishes when you squeeze it.

Below the mantle is the **core**. The core is primarily composed of iron and nickel and is subdivided into the outer core and the inner core. The **outer core** is molten liquid, while the **inner core** is a solid ball. The temperature of the inner core is estimated to be roughly 5,700° Celsius (10,292° Fahrenheit), which is the same as the surface temperature of the sun.

The extreme heat of the inner layers of the earth creates convection currents in the viscous asthenosphere. Lower portions of the asthenosphere slowly heat, expand, rise upward, and then slowly cool and sink. **Plates**, or pieces of the lithosphere, are affected by these currents. The currents pull along the undersurfaces of the lithosphere's various pieces, causing them to slowly move.

Additionally, the cool crust is more solid and dense than the layers below it. This causes lithosphere plates to slowly sink and to melt into the lower layers. This sinking does not happen all at once but occurs gradually along one of the edges of a plate. As one edge sinks, a small gap is created along the opposite edge, and, through this gap, molten rock is free to escape. This rock then cools, solidifies, and adds its own mass to the edge of the plate. This cycle continues and, ever so slowly, the newly-erupted rock will eventually progress to the sinking edge and be melted once more.

The movement of the lithosphere is called **plate tectonics**, and it provides the explanation for the drifting continents that Alfred Wegener theorized. Plate tectonics has now been verified in a variety of ways. The discovery of mid-ocean ridges revealed plate edges where new crust was being formed. Studies of mid-ocean ridges show that the crustal rocks on either side of the ridges have indeed been slowly drifting apart. Advanced global positioning satellite tracking systems can detect the ongoing movements of the continents and even record their speeds.

As plates move, they sometimes come into conflict and collide. The boundary where two plates collide can be a place where tremendous pressure builds. Such plate boundaries are often sites of sudden pressure releases, in the form of volcanoes and earthquakes, and/or of gradual pressure releases, which can slowly build mountain ranges.

## THE WORLD OF DINOSAURS

**D**ue to the actions of plate tectonics, Earth during the Age of Dinosaurs was different from what it is today. By the end of the Permian Period and the beginning of the Triassic Period, all the world's continents had collided together and formed the single supercontinent **Pangaea**. This meant that all the world's oceans were also one. We call this single superocean **Panthalassa**.

Because Pangaea was a single unbroken landmass, the first

dinosaurs that appeared during the Triassic were able to spread across the entire planet, with no major sea barriers standing in their way. For this reason, during the Late Triassic and Early Jurassic, dinosaurs all across the world were fairly similar. Prosauropods and small theropods similar to *Coelophysis* are found worldwide.

During the Jurassic, Pangaea began to slowly split into two smaller supercontinents. **Laurasia** was the northern of the two and was composed of what we today call North America, Europe, and Asia (minus India). **Gondwana** was the southern of the two and was composed of what we today call Antarctica, Australia, Africa, Madagascar, India, and South America. Later, Laurasia and Gondwana also split into smaller continents, but the continents did not assume their modern positions until long after the dinosaur extinction. As the continents drifted apart, so, too, did populations of dinosaurs. Some groups went extinct in Laurasia or Gondwana, and some groups diversified.

By the Cretaceous, there were significant regional differences among the world's dinosaurs. In Laurasia, sauropods were no longer as abundant as they once were, but hadrosaurs and ceratopsians were common. In Gondwana, sauropods did not decline, and carcharodontosaurids and abelisaurs were the dominant theropods. By the Late Cretaceous, Gondwana had begun to break apart into its constituent continents, but Antarctica and Australia remained connected until very near the end of the Cretaceous.

# PALEOCLIMATES

The average global climate was also different during the Age of Dinosaurs. Temperatures were, on average, much higher. This warmer global climate was largely caused by high volcanic activity, which released large quantities of carbon dioxide into the atmosphere. Carbon dioxide is a greenhouse gas that holds in solar heat. The concentration of all Earth's landmasses in only one or two supercontinents may have also been a factor that contributed to the high average temperatures, because it affected the circulation of both air and water currents through the polar regions.

Ocean currents are extremely important to distributing and moving heat from one part of the earth to another. Today, the Gulf Stream is an ocean current that flows from the Gulf of Mexico to the western coast of Europe. Gulf Stream water is warmed in the Gulf of Mexico, and this heat is carried north as it flows.

The Gulf Stream makes many European countries much warmer than places of similar latitudes on the other side of the Atlantic. This is why the maritime provinces of Canada are frigid places in winter, while Italy and Spain rarely see snow. Today, a strong and cold ocean current encircles much of Antarctica, and this current helps keep Antarctica cold. However, 70 million years ago, when Antarctica was still attached to Australia, this current had to go up and around Australia. As the current moved through more equatorial areas, it became warmer, and then, as it flowed back down, it carried this heat to Antarctica.

As a consequence of the high global temperatures, there were no polar ice caps or glaciers during the Mesozoic Era. Antarctica and Australia were located within the Antarctic Circle, and parts of North America were located above the Arctic Circle (North America was actually located further north than it is today). The discovery of lush plant fossils in polar regions indicates that the climate there must have been much warmer than today.

Although it was once assumed that dinosaurs were limited to warm tropical climates, it is now known that many varieties of dinosaurs thrived in polar regions. The Early Jurassic theropod *Cryolophosaurus*, and the prosauropod *Glacialisaurus*, were discovered in Antarctica, along the Transantarctic Mountains. The small ornithopod *Laeallynasaura* inhabited what were then polar forests in the Early Cretaceous of Australia. The Late Cretaceous of Alaska was home to a diverse assemblage of dinosaurs, including the hadrosaur *Edmontosaurus*, the tyrannosaurid *Albertosaurus*, and the ceratopsian *Pachyrhinosaurus*.

**Figure 10.2.** Dinosaurs of the far North: the hadrosaur *Edmontosaurus*. (Figure by Joy Ang)

**Figure 10.3.** Dinosaurs of the far North: the tyrannosaurid *Albertosaurus*. (Figure by Joy Ang)

**Figure 10.4.** Dinosaurs of the far North: the ceratopsian *Pachyrhinosaurus*. (Figure by Joy Ang)

Although the climate was warmer and supported lush polar forests during the Age of Dinosaurs, the polar regions would still have experienced periods of reduced sunlight or total darkness, as they do today. Presumably, photosynthesis would have been reduced during these darker periods. There is debate over whether polar dinosaurs overwintered at the poles or migrated to lower latitudes.

# HIGH SEAS

Today, polar ice caps and glaciers hold large quantities of water, but during the Mesozoic Era, this water was liquid and contributed significantly to high global sea levels. The warm climate also made the average global ocean temperature higher, which led to thermal expansion, causing the world's oceans to further swell and rise. During the Mesozoic, sea levels were up to 250 meters (820.21 feet) higher than they are today. This resulted in the flooding of vast regions and splitting of areas that are now connected into isolated islands.

For example, during the Late Cretaceous, much of the interior of North America was covered by a massive inland sea. This vast waterway spread from the Arctic Ocean to the Gulf of Mexico. At various points during the Mesozoic, North America was subdivided into two separate island subcontinents—Laramidia in the west, from which the majority of North American dinosaur species are known, and Appalachia in the east, from which far fewer dinosaurs are currently known. Because of this ancient sea, Mesozoic marine fossils can be found in now high and dry areas of the American Midwest and Central Canada.

## DINOSAUR DIVERSITY

As this chapter previously discussed, because Pangaea was a single unbroken landmass, the first dinosaurs that appeared during the Triassic were able to spread across the entire planet, with no major sea barriers standing in their way. For this reason, during the Late Triassic and Early Jurassic, dinosaurs all across the world were fairly similar. That changed as Pangaea split apart. In the context of drifting continents, rising and falling seas, and changing climates, different dinosaur populations found themselves isolated and facing different environmental challenges. These challenges pushed dinosaurs to diversify and develop a wide array of adaptations.

In Chapter 1, you were introduced to the major dinosaur groups, but that was only a brief overview. Now that we have also covered plate tectonics and the geologic time scale, it is possible to take a closer look at each branch of the dinosaur family tree and chart its origin and evolution across an ever-changing global landscape.

# THE JURASSIC WORLD OF GIANTS

The first sauropods appeared very late in the Triassic, alongside their prosauropod relatives. During the Early Jurassic, while all the continents were still connected, sauropods rose to new heights, surpassing prosauropods in both abundance and body-size. Among the thriving Jurassic long-necks were the **diplodocids**.

Even compared with other sauropods, most diplodocids have extremely long necks. They are also characterized by front legs that are much shorter than their hind legs, and by their unusual faces. The skull of a diplodocid is elongated and resembles the general shape of a horse's or a deer's. Diplodocid teeth are simple, peg-like, and positioned only at the front of the mouth, not on the sides. They are nipping teeth—good for cropping off leaves and other tender growth. Most diplodocids probably used their long necks to reach high into the trees and used their simple front teeth to nip off the choicest foliage. Some paleontologists also suspect that diplodocids were able to extend their treetop reach even further by rearing backward and standing on only their hind legs while bracing themselves with their tails.

In addition to possibly making use of their tails in tripodal feeding stances, diplodocids also likely employed them as defensive weapons. The end of a diplodocid's tail was made up of a long chain of elongated caudal vertebrae with flexible articulations. With powerful muscles at the base of the hips for swinging, a lash from a diplodocid tail would pack a lethal wallop. In fact, mathematical modeling of diplodocid tails suggests that, at full power, the tail tips could have been cracked like a bullwhip—creating a sonic boom!

Diplodocids shared their Jurassic world with another group of sauropods, called the **macronarians**. Macronarians do not have the whip-tails of diplodocids. Their bodies are generally more robust, and their front legs are usually not noticeably shorter than their back legs. In fact, in macronarians like *Brachiosaurus* and *Giraffatitan*, the front legs were much longer than the back legs.

Macronarians would have had a harder time rearing up and standing tripodally. But most macronarians still have the long necks characteristic of sauropods, and they, too, filled the **ecological niche** of high browsers.

A niche is an animal's way of life. Think of it as the animal's job in the ecosystem—it is how a particular species makes its living, or what it must do to survive. With so many kinds of sauropods living side by side in the Late Jurassic, it might seem like the niche of high browser would have been filled many times over and that sauropods would have faced excessive competition for their food resources. But that was not the case.

Consider the macronarian *Camarasaurus* and the diplodocid *Diplodocus* (the namesake of the group). The bones of both these sauropods have been found side by side in many fossil quarries from the Morrison Formation, in the American West. In comparison, the snout of *Camarasaurus* is much shorter, and its teeth are not limited to the front. In fact, the teeth of *Camarasaurus* line the entire jaw, and the individual teeth are not simple pegs, they are broad, robust, and look like the heads of spoons. While *Diplodocus* has the mouth of a selective nipper, *Camarasaurus* has the mouth of a powerful muncher.

**Figure 10.5.** The skull of *Camarasaurus*, on display at the Royal Tyrrell Museum. (Figure by Amanda Kelley)

**Figure 10.6.** The skull of *Diplodocus*. (Figure by Amanda Kelley)

Diplodocids were adapted to reach high and prune off the most delectable Jurassic foliage, while macronarians were less-picky eaters. Macronarians could crunch much harder, even woody, vegetation, and they could eat what the diplodocids left behind. Thus, these two rather similar animals avoided direct competition for food resources. This is an example of a common ecological phenomenon called **niche partitioning**.

If two species try to occupy the exact same niche, they will compete with each other. Although one species may ultimately outcompete the other, before the loser becomes extinct, both species will suffer from the competition. Thus, competing species always have an evolutionary pressure to adaptively diverge from each other and to become specialized.

Grazing and browsing in the shadows of sauropods were a variety of smaller Jurassic herbivores. Among them were the thyreophorans. By far the most well-known of the Jurassic thyreophorans was *Stegosaurus*. Stegosaurs were a widespread group of thyreophorans in the Jurassic, and their fossils have been found in Africa, Asia, Europe, and North America.

Another group of common Jurassic ornithischians was the ornithopods. Small ornithopods had long legs and appear to have made up for their diminutive size with speed, earning themselves the nickname "Jurassic gazelle." A few Late Jurassic ornithopods obtained greater size, like *Camptosaurus*—an early iguanodont.

The Jurassic niche of big predator was filled by an array of carnivorous dinosaurs. There were giant ceratosaurs and megalosauroids, both ancient lineages of theropods. But the Late Jurassic was a time of predatory change. A new group of big carnivorous dinosaurs had evolved, and they were mounting an ecological takeover.

The **allosauroids** were different from the big predators that had come before them. Allosauroids have vertebrae that interlock more rigidly, so their spines were held stiffer. Their legs are also proportionately longer, suggesting that they were faster than both megalosaurids and ceratosaurids. The allosauroid *Allosaurus* is known from more fossil skeletons than any other big theropod dinosaur, and it was clearly among the most successful of the Late Jurassic's predators.

**Figure 10.7.** The Jurassic allosauroid *Allosaurus*. (Figure by Joy Ang and Veronica Krawcewicz)

Of course, not all the Jurassic carnivores were big. The chicken-sized theropod *Compsognathus* is among the smallest of all known dinosaurs. Like *Allosaurus*, *Compsognathus* has a more rigid spine and long legs, but it belongs to another theropod group—it is a coelurosaur. **Coelurosaurs** are characterized by a long series of sacral vertebrae, narrow hands, and tails with back halves that are skinny, stiff, and lightweight.

**Figure 10.8.** The Jurassic coelurosaur *Compsognathus*. (Figure by Joy Ang and Veronica Krawcewicz)

The Late Jurassic must have been a terrifying time for coelurosaurs. They were predators, but at their size, they could have only hunted for small mammals, tiny lizard-like reptiles, insects, or, in at least a few piscivorous species, small fish. They could have been crushed underfoot by sauropods, thyreophorans, or big ornithopods, and they were certainly an occasional snack for allosauroids and the other theropods, not to mention the various non-dinosaur predators—like Jurassic crocodyliforms. But in the Jurassic, it was the coelurosaurs that spawned the dinosaurs' greatest success: birds. And in the Cretaceous, some coelurosaurs would evolve their way to the very top of the food chain.

# CRETACEOUS GONDWANA

Between the Late Jurassic and the Late Cretaceous, nearly 50 million years of evolution and plate tectonics changed the dinosaur world dramatically. The supercontinents Gondwana and Laurasia split from one another. Some dinosaur groups thrived and multiplied, while others declined and became extinct.

Still, in Late Cretaceous Gondwana, one feature of the Jurassic remained: the sauropods continued to rule supreme as the dominant large herbivores. However, they were not the same sauropods that populated the Jurassic. The diplodocids lumbered off their mortal coil shortly after the Cretaceous began. Some of their close relatives (other diplodocoids) persisted, but they were less common and much smaller than the giants of the Jurassic. On the other hand, the macronarians continued, but by the Late Cretaceous, there were no camarasaurs or brachiosaurs. Instead, a new type of macronarian dominated: the titanosaurs.

**Titanosaurs** are the most robust of all sauropods. Their chests are broad, and their hips are wide. Their hindlimbs are spaced far apart, giving them a very stable base. Many titanosaurs had osteoderms. Titanosaurs ranged in size, but among their ranks were animals like *Argentinosaurus*—a sauropod that has been estimated to weigh over 100 tons, making it the largest creature to ever walk the earth.

With armor and sheer size to protect them, titanosaurs were not easy prey. But one group of Late Cretaceous theropods may have been specialized giant-slayers. The **carcharodontosaurs** are named for the shape of their teeth, which resemble those of the *Carcharodon*—the great white shark. Carcharodontosaurs are a type of allosauroid, so they are descendants of the big theropods that first rose to prominence in the Late Jurassic. However, carcharodontosaurs differ from older allosauroids in a number of ways. Most noticeably, carcharodontosaurs have bigger heads, with longer jaws.

Some paleontologists speculate that, to kill titanosaurs, carcharodontosaurs used their mouths for more than simple biting. With strong muscles connecting their heads to their necks, the shark-toothed dinosaurs may have opened their jaws wide agape and swung their entire heads in a downward arc against a sauropod's flank. Using this motion, their upper jaws would have acted like giant meat cleavers and sliced off huge chunks of flesh.

As Late Cretaceous titanosaurs got bigger, so did carcharodontosaurs. The largest of all was the South American *Giganotosaurus*. At over 13 meters (42.65 feet) in length, *Giganotosaurus* even outsized *Tyrannosaurus rex*.

**Figure 10.9.** The carcharodontosaur *Giganotosaurus*. (Figure by Joy Ang and Veronica Krawcewicz)

There was room for more than one kind of big carnivore in the Late Cretaceous of South America. **Abelisaurs**, like the famous horned species *Carnotaurus*, were the last survivors of the ceratosaur lineage, and some grew to over 8 meters (26.25 feet) in length. In the Cretaceous, the group was strictly limited to Gondwana, but they evidently thrived there, as abelisaur fossils have been found throughout the southern hemisphere.

**Figure 10.10.** The abelisaur *Carnotaurus*. (Figure by Joy Ang and Veronica Krawcewicz)

Living alongside carcharodontosaurs must have been tough, and the need for ecological niche partitioning drove abelisaurs to adapt a strikingly different morphology. While carcharodontosaurs have long jaws with big teeth, abelisaurs have short muzzles and proportionately tiny teeth. While carcharodontosaurs tended to have powerful forearms with large hooked claws, abelisaurs had ridiculously short and stubby arms, with small claws. And, while carcharodontosaurs likely preyed on huge titanosaurs, abelisaurs are thought by many paleontologists to have hunted the smaller species of titanosaurs and other less-daunting herbivores.

*Carnotaurus* shows off another trait common to most abelisaurs: a rugose, or wrinkly, bone texture on the skull and large cranial ornamentations. The rugosity suggests that, in life, the faces of abelisaurs were covered with tough keratinous pads. The cranial ornamentations, which took the form of bony horns and large lumps, were probably sexual display structures. It has also been suggested, in the case of *Carnotaurus*, that the horns might have been used in agonistic head to head shoving and jousting competitions between males over mating rights and territory.

# CRETACEOUS LAURASIA

While titanosaurs were abundant and diverse in Gondwana, they were far less common in Late Cretaceous Laurasia. In what is now Asia, titanosaurs were important but comparatively rare components of the ecosystem. In North America, only a handful of titanosaur species are known. This relative underabundance of sauropods in the north is one of the biggest differences between the Late Cretaceous fauna of Laurasia and Gondwana and meant that northern herbivorous niches were filled by other kinds of plant-eaters.

Although stegosaurs never made it to the Late Cretaceous, another group of thyreophorans did: the ankylosaurs. In Laurasia, ankylosaurs split into two major groups. The **ankylosaurids** are the ankylosaurs with the famous tail clubs. Ankylosaurids also typically have large backward-pointing horns at the rear of their skulls and a short rounded snout at the front.

**Nodosaurids** are the second major group of ankylosaurs. They lacked tail clubs, but some have offensive weapons at the other end, in the form of large osteoderm spikes that project outward from over their shoulders. Nodosaurids do not generally have the big skull horns of ankylosaurids, and their snouts are significantly narrower and more elongated.

After their start in the Jurassic, with the likes of *Camptosaurus*, iguanodont ornithopods thrived and became common across the globe during the Early Cretaceous. In Laurasia, a new kind of iguanodont evolved: the hadrosaurs. Hadrosaurs flourished in the Late Cretaceous and quickly became the northern hemisphere's most successful herbivorous dinosaurs.

We have more fossils of hadrosaurs and know more about hadrosaur biology than any other major dinosaur group. The success of these duck-billed dinosaurs was thanks in large part to their sophisticated dental batteries, which let them chew through Mesozoic plants like nothing else ever had. Hadrosaurs were most abundant in Laurasia, but the group did manage to spread to

Gondwana, and hadrosaur fossils have been found as far south as Antarctica.

Among advanced hadrosaurs, there are two major groups. The **lambeosaurine** hadrosaurs had a big crest on their heads, which can be thought of as a kind of horn. Not a "horn" like on the face of a rhinoceros or *Triceratops*, but rather a "horn" as in a musical instrument. Inside a lambeosaurine crest is a complex and hollow nasal passageway. Blowing air through this passage and then out the nostrils would have amplified the dinosaur's calls.

The hollow crests of lambeosaurines come in a variety of sizes and shapes. This not only gave each species a unique appearance but also a unique sound. Studies of hadrosaur ears indicate that the group had excellent hearing. Like the unique calls and songs of modern birds, the species-specific sounds that hadrosaurs generated with their cranial instruments could have been used to court mates, establish territories, or communicate a variety of messages to members of their own kind.

The second major group of advanced hadrosaurs is the **hadrosaurines** (sometimes called the saurolophines). From the skulls of hadrosaurines, it is clear that they do not have the complex sound-amplifying crests of the lambeosaurines. However, some hadrosaurines do still have crests. For instance, the hadrosaurine *Saurolophus* had a prominent, but solid, bony crest. Recently, a fossil mummy specimen of the hadrosaurine *Edmontosaurus* was discovered in Alberta, Canada, with a big fleshy crest, like the comb of a rooster, preserved on the top of its head. This specimen revealed that at least some hadrosaurines had large crests, even though their skulls provide no record of them.

Running a close second to hadrosaurs, in terms of Laurasian diversity and success, was another group of herbivorous dinosaurs: the **marginocephalians**. That name literally means "fringe heads" and refers to an overhanging lip of bone at the back margin of the skull. Pachycephalosaurs are one of the two major groups within the marginocephalians. The other is the ceratopsians.

The first ceratopsians are a far cry from the later and famous forms, like *Triceratops*. Primitive ceratopsians, like *Psittaco-saurus*, are small and bipedal dinosaurs, but they still show a few family resemblances. Like all ceratopsians, they have large beaks and small jugal cheek-horns projecting from the sides of their face.

**Figure 10.11.** The early ceratopsian *Psittacosaurus*. (Figure by Joy Ang and Veronica Krawcewicz)

*Psittacosaurus* and other primitive ceratopsians are abundantly known from Asia. However, the larger advanced ceratopsians, like *Triceratops*, are mostly known from North America. This kind of diversity within Laurasia was possible during the Late Cretaceous, because by this time, the continent was itself beginning to break apart.

For much of the Cretaceous, the allosauroids were the top predators throughout Laurasia, just as they were in Gondwana. But they were rivaled by another theropod group. The coelurosaurs came a long way from the small and humble forms of the Jurassic. Coelurosaurs blossomed into the most diverse of all theropod groups, and they gave rise to the most infamous of all predatory dinosaurs: the tyrannosaurs.

By the end of the Cretaceous, tyrannosaurs came to dominate the niche of alpha predators throughout Laurasia. As coelurosaurs, the tyrannosaurs did not inherit their position from their ancestral lineage. Rather, they ousted the allosauroids, just as the allosauroids had ousted the large megalosauroids and ceratosaurs back in the Jurassic. To some extent, tyrannosaurs achieved their success by taking to a further extreme the same adaptations that had once set the allosauroids apart from their competitors; that is, tyrannosaurs evolved even longer legs and a much stiffer vertebral column. However, tyrannosaurs also evolved massive skulls with tremendous jaw muscles.

The earliest tyrannosauroids, like the Asian species *Dilong* and *Guanlong*, or the European species *Eotyrannus*, have normal head and body proportions, and they look similar to most other small coelurosaurs. But as tyrannosaurs evolved, they grew in absolute size and in the relative size of their heads. These big heads added weight to the front half of their bodies. To compensate, tyrannosaurs reduced weight by shrinking the size of their arms and hands, culminating in the last and largest of all tyrannosaurs—*Tyrannosaurus rex*.

**Figure 10.12.** The early tyrannosauroid *Eotyrannus*. (Figure by Joy Ang and Veronica Krawcewicz)

Not all Cretaceous coelurosaurs were huge and predatory. **Ornithomimids** are a kind of coelurosaur that evolved a body plan similar to that of a modern ostrich or emu but with long clawed forelimbs and a large tail. Like ostriches, ornithomimids have beaks, have long legs for fast sprinting, and were probably mostly herbivorous.

Another group of coelurosaurs developed a highly specialized wrist bone called a **semilunate carpal**. These crescent-shaped bones allowed the hand to be folded backward at a sharp angle, and the dinosaurs that possess them are called the **maniraptorans**. Birds are one group of maniraptorans, and the semilunate carpals of birds allow them to delicately fold their wings when not flying.

As close relatives of birds, the sickle-clawed deinonychosaurs are also maniraptorans, and so are the **oviraptorosaurs**. Like ornithomimids, oviraptorosaurs are a group of theropods that adapted to a mostly vegetarian life and lost their teeth in favor of large beaks. Many oviraptorosaurs had cranial crests and fans of feathers on the ends of their tails.

**Figure 10.13.** The deinonychosaur *Hesperonychus*. (Figure by Joy Ang and Veronica Krawcewicz)

**Figure 10.14.** The oviraptorosaur *Citipati*. (Figure by Joy Ang)

Perhaps the strangest of all Laurasian coelurosaurs are the **therizinosaurs**. Therizinosaurs are also probably the most confusing. The first therizinosaur fossil to be found was a huge claw, over 60 centimeters (23.62 inches) long. Paleontologists had no particular reason to think the claw belonged to a dinosaur. It certainly did not look like the claw of any dinosaur that had ever been found before. Instead, it was mistaken for the claw of a giant turtle.

Even after more fossils were found, and it was clear that therizinosaurs were dinosaurs, not turtles, no one was quite sure what kind of dinosaurs they were. Therizinosaurs have small skulls on the end of long necks, and they have hind feet with four forward-pointing toes, so some paleontologists thought they might be prosauropods. Therizinosaurs also have a backward-directed pubis and jaws with small herbivorous teeth in the back and a beak in the front, so some researchers classified them as ornithischians. Several fossil skeleton discoveries and a lot of research later, paleontologists are now in agreement that therizinosaurs are maniraptoran theropods (close relatives of oviraptorosaurs)—with the semilunate carpals to prove it.

**Figure 10.15.** The therizinosaur *Therizinosaurus*. (Figure by Joy Ang and Veronica Krawcewicz)

But the controversies and uncertainties surrounding therizinosaurs are not over. Paleontologists are currently debating what exactly therizinosaurs ate and what their ecological niche was. For instance, Dr. Philip Currie (head of the University of Alberta dinosaur research program) speculates that therizinosaurs were piscivores and used their long arms and claws for spearing fish. If you asked paleontologist W. Scott Persons, he would argue that the short beaks and teeth and the backward-directed pubic bone of therizinosaurs indicate a mostly herbivorous diet and that these ungainly creatures needed those wicked claws for defense. Other paleontologists think of therizinosaurs as giant anteaters that specialized in an insectivorous diet and used their claws for tearing into termite mounds and anthills.

## WORLDS COLLIDE

The movement of the continents does not always lead to geographic isolation and evolutionary diversification. Sometimes continents drift together, not apart. When that happens, species from each of the converging continents are free to spread to the other. This can lead to very similar flora and fauna on both landmasses.

For instance, during the Late Cretaceous, there is good evidence of a direct land connection between Northwestern North America and East Asia. Dinosaur species discovered in Alberta's Dinosaur Provincial Park have very close cousins in the Late Cretaceous beds of Mongolia's Gobi Desert. Asia and North America were probably intermittently connected via Alaska at this time. When plate tectonics or changes in sea level bring landmasses together, and animals migrate between them, we call this phenomenon a **faunal interchange**.

## CHAPTER 11

# Dinosaur Origins

**LEARNING OBJECTIVE FOR CHAPTER 11:** Understand the early evolution of dinosaurs from non-dinosaurian archosaurs.

- **Learning Objective 11.1:** Recognize features that differentiate anapsids, synapsids, and diapsids.
- **Learning Objective 11.2:** Recognize features that differentiate lepidosauromorph diapsids and archosaur diapsids.
- **Learning Objective 11.3:** Recognize features that differentiate pseudosuchian archosaurs from avemetatarsalian archosaurs.
- **Learning Objective 11.4:** Understand the vertebrate fauna characteristic of the Permian Period and the changes in vertebrate community structure at the Permian-Triassic boundary.
- **Learning Objective 11.5:** Understand the vertebrate fauna characteristic of the Triassic Period and the changes in vertebrate community structure at the Triassic-Jurassic boundary.

# SKULL FENESTRAE

Recall that fenestrae are additional openings in the skull that do not house sensory organs. Usually, fenestrae provide an open area for large muscles to fill. The number and arrangement of fenestrae are key characters that are used to help classify amniotes into their major lineages.

Amniotes that completely lack fenestrae are called **anapsids**. Anapsids were common earlier in the history of amniotes. **Synapsids** are amniotes with one fenestra on each lateral side of their skull. All mammals are synapsids, and so were our close reptilian ancestors, like the famous sail-backed synapsid ***Dimetrodon***. Although it is commonly misidentified as a dinosaur, *Dimetrodon* is more closely related to you and me than it is to any dinosaur. *Dimetrodon* lived during the Permian Period, so it was millions of years older than the first dinosaurs.

**Figure 11.1.** Skeleton of *Dimetrodon*, in the Royal Tyrrell Museum of Palaeontology. (Figure by Amanda Kelley)

Amniotes with one set of fenestrae on the lateral sides of their skulls and one set on the top surfaces of their skulls are called **diapsids**. Diapsids are further subdivided into two groups, again based on fenestrae. **Lepidosauromorphs** are diapsids with no additional fenestrae. Modern lepidosauromorphs include lizards, snakes, and tuataras. **Archosaurs** are diapsids with an additional fenestra in front of each orbit (the **antorbital fenestra**) and an additional fenestra on the rear of the lower jaw (the **mandibular fenestra**). Crocodilians, birds, dinosaurs, and the extinct flying reptiles called pterosaurs are all archosaurs.

Note that some lineages of archosaurs, such as modern crocodilians, have secondarily lost their antorbital fenestra, and some, like the pterosaurs, secondarily lost their mandibular fenestra. This does not mean that crocodilians or pterosaurs lose their status as archosaurs, because "archosaur" is a name applied to the evolutionary lineage. As long as the ancestors of crocodilians and pterosaurs had the characters that define an archosaur (and they did), crocodilians, pterosaurs, and all other such descendants will be classified as part of this evolutionary group.

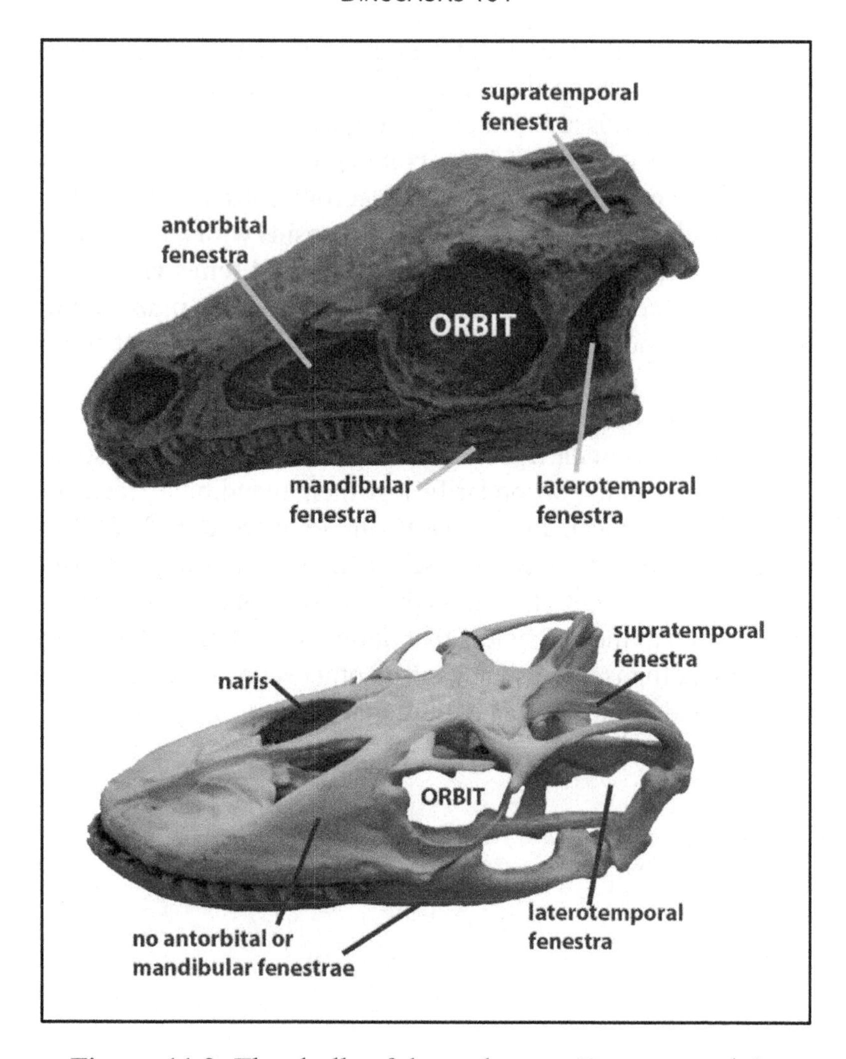


**Figure 11.2.** The skulls of the archosaur *Eoraptor* and the lepidosauromorph *Varanus komodoensis* (Komodo dragon). (Figure by Victoria Arbour)

# ANKLES

Dinosaurs, pterosaurs, and a few of their close relatives belong to a special group of archosaurs and are known as **avemetatarsalians**. Avemetatarsalians are characterized by having ankles that flex like a hinge, while other archosaurs have ankles that rotate like a ball-and-socket. This adaptation gave avemetatarsalians stiffer ankles, which were better able to safely support their weight while running and were better suited to locomotion on upright (non-sprawling) limbs. On the basis of their ankles and a few other characters, the archosaurs are divided into two main lineages: the hinge-ankled **avemetatarsalians** and the ball-and-socket-ankled **pseudosuchians**, which include living crocodilians and many extinct groups.

# THE PERMIAN—A PROLOGUE TO DINOSAURS

In the Permian Period (299 to 252 million years ago), all the world's landmass was part of the supercontinent Pangaea. This single continent had an arid interior, with rapidly fluctuating temperatures and climates. The first group of amniotes to evolve large body size and to dominate the ecological roles of mega-herbivores and mega-carnivores across Pangaea was not the dinosaurs, it was the synapsids—our own lineage! Reptile-like synapsids, including *Dimetrodon*, became common and thrived for millions of years.

Gradually, these early synapsids became more mammal-like. Late in the Permian, large saber-toothed synapsids called **gorgonopsids** were the top predators, and synapsids like the tusked **dicynodonts** were the top herbivores. There was a diverse array of small and medium-sized synapsids, including the cynodonts. **Cynodonts** would go on to evolve into true mammals, and the early forms looked a little like short-legged dogs. Then, 252 million years ago, disaster struck the world of the synapsids.

**Figure 11.3.** The gorgonopsid *Gorgonops*. (Figure by Rachelle Bugeaud and Veronica Krawcewicz)

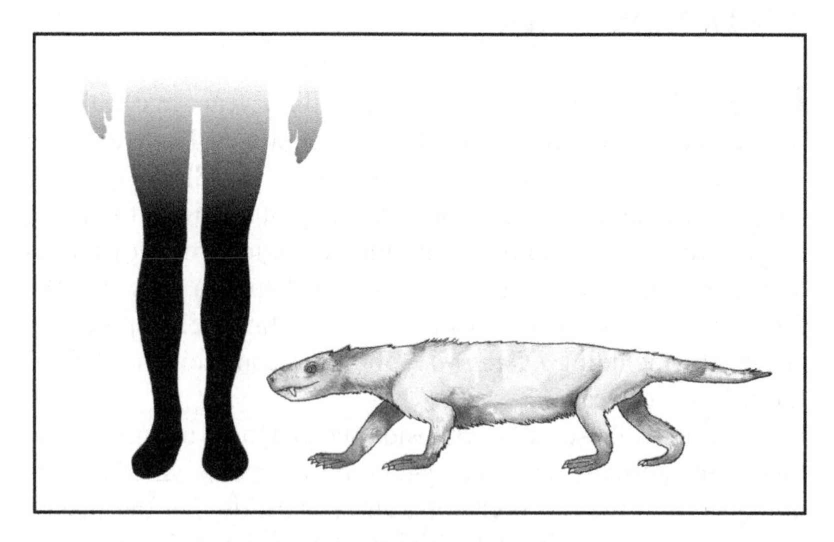

**Figure 11.4.** The cynodont *Thrinaxodon*. (Figure by Rachelle Bugeaud)

The end-Permian mass extinction was the most devastating extinction event in the history of life. Estimates of exactly how many species went extinct vary. But paleontologists agree that about 70% of all terrestrial vertebrate species and 90–95% of all marine species died in the event.

The cause (or causes) of the end-Permian mass extinction remains uncertain. Huge lava deposits, known as the Siberian Traps, formed at this time, and the massive volcanic eruptions that created these deposits must have released large quantities of volcanic gases and clouds of ash and dust into the atmosphere. These volcanic clouds may have blocked out sunlight and led to cooling temperatures. When they poured into the ocean, these same lava flows may have also triggered a release of large amounts of methane gas. The released methane gas would have insulated the planet's atmosphere and have led to a sudden global temperature increase.

Finally, the extinction may have been brought about by a comet or meteorite impact, although a crater from such an impact has yet to be found. Whatever the cause, the end-Permian mass extinction was the single greatest extinction event ever, and it took millions of years for Earth's ecosystems to recover.

## THE TRIASSIC'S NEW WORLD ORDER

The synapsids had been cut down in their evolutionary prime, and this left vacant the ecological roles that they had previously filled. At first, the synapsid lineages that had managed to survive the extinction slowly rebounded, and some groups re-evolved large body sizes and reassumed their roles as top predators and herbivores. Cynodonts and dicynodonts were among the synapsids that succeeded in rebounding, and it is during the Triassic that the first true mammals appeared. However, at the same time, a new group of diapsids, the archosaurs, also began to diversify and grow. Gradually, large archosaurs became more abundant, while large synapsids became less abundant.

The first widely successful group of archosaurs was a lineage that would later go on to evolve into modern crocodilians. These crocodile-line archosaurs are called **pseudosuchians**. The pseudosuchians of the Triassic include the often huge and slender-snouted phytosaurs, which were semiaquatic predators like their distant crocodile relatives; the heavily armored and herbivorous aetosaurs; the rauisuchids and prestosuchians, which were terrestrial predators with upright limb posture; and the poposauroids, some of which were sail-backs and demonstrate convergent evolution with the earlier *Dimetrodon*.

**Figure 11.5.** The phytosaur *Rutiodon*. (Figure by Rachelle Bugeaud and Veronica Krawcewicz)

**Figure 11.6.** The aetosaur *Desmatosuchus*. (Figure by Rachelle Bugeaud)

**Figure 11.7.** The prestosuchian *Prestosuchus*. (Figure by Rachelle Bugeaud and Veronica Krawcewicz)

**Figure 11.8.** The poposauroid *Arizonasaurus*. (Figure by Rachelle Bugeaud and Veronica Krawcewicz)

Where were the dinosaurs? The oldest record of dinosaur-like archosaurs comes from footprints that have been dated at roughly 250 million years old. The earliest dinosaur-like archosaurs were small and bipedal and looked a lot like the true dinosaurs, but they lacked some of the specializations that characterize true dinosaurs, such as a hip socket with a hole through it—for this reason, we call these animals **dinosauromorphs**.

The best record of early dinosaur bones comes from 228-million-year-old fossil beds of Argentina. *Eoraptor*, *Eodromaeus*, *Herrerasaurus*, and *Panphagia* are examples of early carnivorous saurischian dinosaurs, and *Pisanosaurus* is an early herbivorous ornithischian dinosaur. It seems clear that early dinosaurs were more successful and diverse as carnivores than as herbivores. *Eoraptor*, *Eodromaeus*, *Panphagia*, and *Pisanosaurus* are all relatively small (under 1 meter [3.28 feet] in length), but *Herrerasaurus* was significantly larger (comparable in size to a modern tiger). Compared to the many other archosaurs, these early dinosaurs were rare components of their ecosystems.

**Figure 11.9.** The early dinosaur *Eoraptor* was a carnivore and possibly omnivore. (Figure by Rachelle Bugeaud and Veronica Krawcewicz)

**Figure 11.10.** The early herbivorous dinosaur *Pisanosaurus*. (Figure by Rachelle Bugeaud and Veronica Krawcewicz)

**Figure 11.11.** *Herrerasaurus* was among the largest early carnivorous dinosaurs. (Figure by Rachelle Bugeaud and Veronica Krawcewicz)

As the Triassic drew to a close, dinosaurs were gaining ground. *Coelophysis* was a wolf-sized Triassic theropod that has been found in large bone beds in New Mexico. *Coelophysis* appears to have been one of the most common predators of its time and place. Prosauropods evolved late in the Triassic and also were hugely successful. *Plateosaurus* is the best known of the prosauropods and would have weighed more than 3 tons. In the Triassic, prosauropods were record-breakers as the largest herbivores that had ever evolved up to that time.

At the end of the Triassic, another mass extinction event of unknown cause occurred. This extinction was not nearly as severe as the extinction at the end of the Permian. Still, it hit many of the thriving archosaur groups hard . . . but not dinosaurs.

Although some dinosaurs, including many of the prosauropods, were killed off, other dinosaurs not only survived the extinction event but thrived after it. The extinction left several ecological roles vacant, and dinosaurs quickly evolved to fill them. This success, at the time of Pangaea, allowed dinosaurs to spread to the far edges of every continent, without ever having to swim. As Pangaea broke apart, dinosaurs rode the plates, and different dinosaur groups had the opportunity to evolve and diversify in geographic isolation. With the start of the Jurassic, the Age of Dinosaurs had truly begun.

CHAPTER 12

# Dinosaur Extinction

**LEARNING OBJECTIVE FOR CHAPTER 12:** Understand the impact theory of dinosaur extinction and the pattern of extinction at the end-Cretaceous event.

- **Learning Objective 12.1:** Understand the difference between background extinctions and mass extinctions.
- **Learning Objective 12.2:** Recognize which groups of organisms from the Mesozoic Era are no longer alive and which are still alive.
- **Learning Objective 12.3:** Recognize geological features associated with meteorite impacts.
- **Learning Objective 12.4:** Evaluate the evidence for a meteorite impact at the end of the Cretaceous.
- **Learning Objective 12.5:** Understand the approaches to and the problems inherent in attempts to resurrect extinct animals.

# MASS EXTINCTION EVENTS

Species that are still present today are called **extant** species. Species whose members have all died off are called **extinct** species. Naturally, the number of extant species is only a tiny fraction of the huge number of species that are now extinct.

As environments gradually change and species evolve and compete, the extinction of some species is an inevitable result. At any time in the history of life, it is usual for some species to be going extinct. However, certain dramatic environmental changes can trigger the extinction of many species all at nearly the same time and across the entire planet. When such a sudden and global loss of species occurs, it is called a mass extinction event.

Paleontologists generally recognize five major mass extinctions. The end-Ordovician mass extinction affected only marine organisms, but at that time terrestrial organisms had only just begun to evolve. The Late Devonian mass extinction was also largely limited to marine organisms, including some early vertebrate clades. As discussed in Chapter 11, the end-Permian mass extinction saw the largest loss of diversity in all of Earth's history. The end-Triassic mass extinction saw the extinction of most lineages of pseudosuchian archosaurs, as well as many of the synapsids that had survived the end-Permian extinction, and also affected marine life. The last of the "Big Five" extinctions was the end-Cretaceous.

It should be remembered that extinction is not a phenomenon limited to the deep and distant expanses of geologic time. Many species have gone extinct only recently. Perhaps one of the most famous examples, the passenger pigeon, went extinct in 1914, even though there were billions of passenger pigeons only a few decades prior. The Carolina Parakeet, the only species of parrot native to the United States, went extinct in 1918. The thylacine (also called the Tasmanian tiger) went extinct in 1936. Even more recently, following extensive field surveys, the baiji river dolphin was declared extinct in 2006, and the Alaotra Grebe (an African waterfowl) was declared extinct in 2010.

The extinction of all of those species was the result of intense human hunting and habitat loss. Based on the current rate of species extinction, many biologists have argued that the earth is presently in the middle of a sixth mass extinction event. This new mass extinction is being brought about by sudden global climate change and large-scale ecosystem destruction and degradation (the results of human activities).

# THE END-CRETACEOUS EXTINCTION

The end-Cretaceous extinction event occurred roughly 66 million years ago and killed all non-avian dinosaurs. However, dinosaurs were not the only casualties of this extinction. In the oceans, large marine diapsids, called mosasaurs and plesiosaurs, died out, as did many varieties of corals, several forms of plankton, and ammonites (relatives of modern squids and octopi). Pterosaurs went extinct as well. Although birds ultimately survived, many types of Cretaceous birds (including hesperornithiform and enantiornithiform birds) perished. Land plants also lost many species in the extinction, and insect diversity fell.

Mammals, turtles, crocodiles, amphibians, and fish all made it through the end-Cretaceous extinction, although many of the larger species in all these groups did not. Generally, it seems that large animals and photosynthetic organisms were the most likely to die off. Small animals, and particularly those that were semiaquatic, had the best chance of surviving.

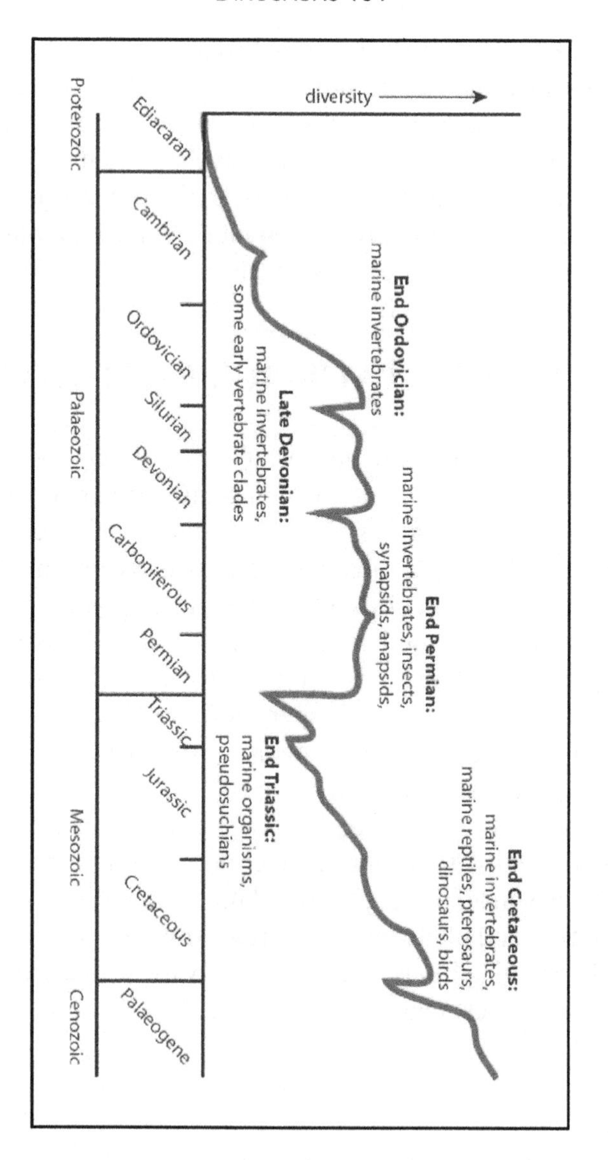


**Figure 12.1.** This figure is modified from a graph of "Family"-level diversity through time by Jack Sepkoski and David Raup published in 1982 and often referred to as the Sepkoski curve. Newer data has refined the graph somewhat, but the overall pattern still stands. (Figure by Victoria Arbour)

Of course, not all groups of animals that survived the end-Cretaceous extinction are still around today. Champsosaurs are a good example: these aquatic diapsids, which look like crocodiles (though the two groups are not closely related), survived the end-Cretaceous mass extinction, only to go extinct during the Early Miocene (about 20 million years ago).

**Figure 12.2.** The skull of *Champsosaurus*. (Figure by Amanda Kelley)

There have been many ideas put forward to try to explain the cause of the end-Cretaceous extinction. Some of these ideas are more plausible than others. It has been suggested that dinosaurs went extinct because small mammals began eating all of the dinosaurs' eggs. Not only does this idea not take into consideration the fact that mammals and dinosaurs evolved at roughly the same time (and, therefore, dinosaurs had been successfully coexisting with small mammals for over 160 million years and laying eggs all the while), but it also fails to explain why so many other kinds of organisms died out at the same time.

It is commonly and incorrectly also thought that the Cretaceous Period was immediately followed by an ice age and that widespread glaciations and freezing temperatures were responsible for the dinosaurs' demise. While average global temperatures did eventually fall after the Cretaceous, this temperature fall was gradual, and it was millions of years before a true ice age resulted.

A mass volcanic outgassing of carbon dioxide and ash plumes has also been suggested as a possible cause of the extinction. This scenario could potentially have affected the global climate enough to have caused the extinction of many kinds of organisms, and there is a record of high volcanic activity in the Deccan Traps of India at the end of the Cretaceous. However, the current prevailing theory for the cause of the end-Cretaceous extinction is more cosmic.

## THE CHICXULUB IMPACT

In 1979, stratigraphers were studying rock layers at the boundary of the Cretaceous and Paleogene periods and noticed a strange thin layer of gray clay. Later, this same gray layer began to be discovered at the Cretaceous-Paleogene boundary all over the world and in very different formations. Close inspection of the gray clay layer revealed that it had high concentrations of iridium. **Iridium** is a rare element on earth, but it is a common component of meteorites.

That was not all: the layer was also rich in tektites and shocked quartz. **Tektites** are tiny pieces of rock that have been melted and then cooled. **Shocked quartz** is a form of the mineral quartz with a unique internal structure that can only be created by exposure to a powerful shockwave. Both tektites and shocked quartz are telltale signs of a meteorite impact. But to spread iridium, tektites, and shocked quartz all across the globe would have required either an enormous shower of large meteorites or . . . a single tremendous meteorite impact. The search was on for a giant crater.

**Figure 12.3.** Tektite specimens, in the University of Alberta Geology Collection. (Figure by W. Scott Persons)

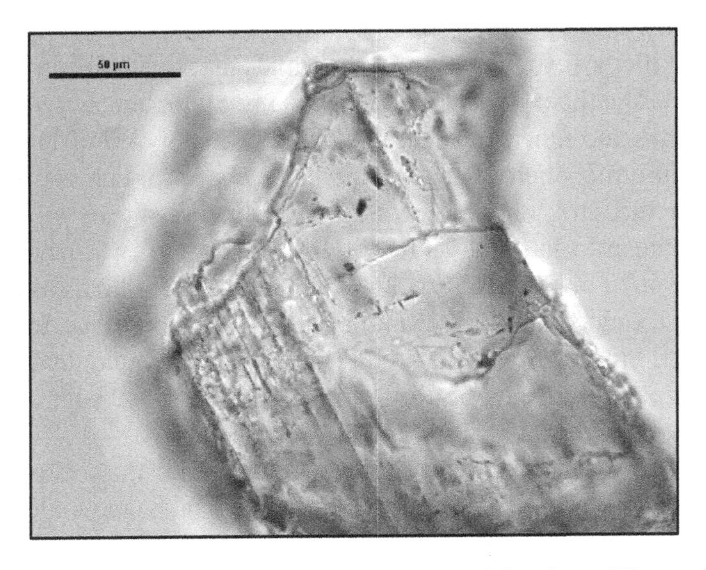


**Figure 12.4.** Shocked quartz under magnification. (Figure by Randy Kofman)

For many years, no such crater was found. Then, geologists working near the town of Chicxulub in Mexico's Yucatán Peninsula noticed a peculiar pattern of **cenotes**, or limestone sinkholes. The cenotes were arranged in a crescent shape many miles long. Each end of the crescent seemed to terminate at an edge of the peninsula.

Investigation revealed that the cenotes were caused by a displaced portion of a limestone layer that had been pushed upward and that the structure did not actually end at the edges of the peninsula. Instead, it continued along the ocean floor and was actually a huge continuous ring over 180 kilometers (111.85 miles) in diameter. Radiometric dating revealed that this massive ring of displaced rock was 66 million years old. The crater made by the meteorite responsible for showering the earth with debris at the end of the Cretaceous had been found. Based on the crater's size, it has been calculated that the meteorite that made it must have been 10 kilometers (6.21 miles) in diameter, larger than Mount Everest.

How could a single meteorite impact, even a massive one, have killed off so many kinds of animals that lived all across the globe? The theory goes like this: The initial impact caused huge tsunamis and sent a great cloud of superheated rock and dust high into the atmosphere. The rock and larger pieces of debris quickly fell to the earth and started wildfires. Smaller pieces of debris next began to fall and, as they fell, were heated by air friction. This rain of hot dust raised global temperatures for hours after the impact and cooked alive animals that were too large to seek shelter. Small animals that could shelter underground, underwater, or perhaps in caves or large tree trunks may have been able to survive this initial heat blast.

Finally, much debris would have remained in the atmosphere for perhaps months or even years. The residual haze would have reduced sunlight, killing many plants and other photosynthetic organisms, with rippling effects up the food chain. The reduced sunlight may also have brought on a sudden drop in global temperatures. Being large active animals with high energy needs and positioned at the top of prehistoric food chains, dinosaurs were highly susceptible to this series of catastrophes, while smaller animals with lower metabolisms were best able to wait the disaster out.

## RESURRECTING EXTINCT SPECIES

Will human eyes ever see a living, breathing *Tyrannosaurus* or *Triceratops*? It is hard to say; but for those of us living here in the twenty-first century, it does not seem likely.

You may be familiar with a certain Hollywood franchise that popularized the science fiction premise of cloning dinosaurs from discoveries of their DNA. Unfortunately, DNA is a delicate substance that quickly breaks down over time. It is extremely unlikely that a complete or nearly-complete DNA strand could ever be preserved (even inside the body of a mosquito stuck in amber) for 66 million years or more. Some recent but controversial research suggests that remains of dinosaur blood vessels and potentially even proteins may be able to survive this long. But such material is still a long way from what would be needed to even consider cloning a dinosaur.

It might be possible to find more recent dinosaur DNA. Remember that birds are one branch of the dinosaur family tree. As such, the DNA of birds contains many of the DNA sequences of their ancestors (but with many of these genes switched off). It has been proposed that a dinosaur could be resurrected by hatching a bird with its advanced DNA sequences turned off and its ancient ancestral sequences turned back on. In this way, perhaps a bird would develop with a long bony tail, teeth, and clawed fingers. But performing such genetic manipulations is well beyond our current understanding and technology. For now, living dinosaurs (except for birds) are things of the past.

# From the Publisher

## Thank You from the Publisher

Van Rye Publishing, LLC ("VRP") sincerely thanks you for your interest in and purchase of this book.

If you enjoyed this book or found it useful, please consider taking a moment to support the authors and get word out to other readers like you by leaving a rating or review of the book at its product page at your favorite online book retailer.

Thank you!

## Resources from the Publisher

Van Rye Publishing, LLC ("VRP") offers the following resources to writers and to readers.

For writers who enjoyed this book or found it useful, please consider having VRP edit, format, or fully publish your own book manuscript. You can find out more, and contact the publisher directly, by visiting VRP's website: www.vanryepublishing.com.

For readers who enjoyed this book or found it useful, please consider signing up to have VRP notify you when books like this one are available at a limited-time discounted price, some as low as $0.99. You can sign up to receive such notifications by visiting the following web address: http://eepurl.com/cERow9.

For anyone who enjoyed this book or found it useful, if you have not already done so, please again consider leaving a rating or review of this book at its product page at your favorite online book retailer. These ratings and reviews are themselves extremely valuable resources for writers and for readers like you. VRP therefore hopes you will please take a moment to share your thoughts about this book with others.

Thank you again!

# About the Authors

**DR. W. SCOTT PERSONS IV** is a paleontologist who holds a bachelor's degree in Geology from Macalester College and a master's degree and PhD in Evolution and Systematics from the University of Alberta, where he was a student of world-renowned paleontologist Dr. Philip J. Currie. Dr. Persons's research focuses primarily on evolutionary arms races between dinosaurian predators and prey, on the biomechanics and evolution of dinosaur locomotion, and on the evolution of dinosaurian feathers. He has taken part in fossil-hunting expeditions throughout the American West, Mongolia, Tanzania, Argentina, and China. Dr. Persons is the author of the book *Dinosaurs of the Alberta Badlands*, and his work has been featured on the *National Geographic* and *Discovery* channels and in *Smithsonian* and *Discover Magazine*.

**DR. PHILIP J. CURRIE** is a full professor and Canada Research Chair of Dinosaur Paleobiology at the University of Alberta. As a world-renowned paleontologist, Dr. Currie's scientific accomplishments have led to a greater understanding of dinosaurs and their significance. He was instrumental in the development of the Royal Tyrrell Museum of Palaeontology and has made major contributions to paleontology on both the Canadian and the world stage through his extensive fieldwork, academic research, writing, and teaching. In addition to the numerous awards and medals Dr. Currie has received, he has published 170 scientific articles, 140 popular articles, and 15 books, has given more than 800 lectures attended by over 50,000 people, and has given more than 1,200 newspaper, magazine, radio, film, and television interviews for articles and programs.

**DR. VICTORIA ARBOUR** is the Curator of Paleontology at the Royal BC Museum in Victoria, British Columbia, Canada. She is a leading expert on the evolution and paleobiology of ankylosaurs, and she studies the evolution of animal weapons, dinosaur biogeography, and the fossil record of British Columbia. She co-created the online course *Dino 101* during her PhD work at the University of Alberta.

**DR. MATTHEW VAVREK** has a PhD from McGill University. After several years of working in education and related fields at museums and institutions across Canada, he returned to school to complete an Education After Degree (BEd) through the University of Alberta's Teacher Education North program. He is currently an instructor at Peace Wapiti Academy and Grande Prairie Regional College in Grande Prairie, Alberta, where he is lucky enough to live 10 minutes away from some of his research sites.

**DR. EVA KOPPELHUS** is a professor and curator at the University of Alberta. She researches paleobotany (fossil plants) and palynology (fossilized spores and pollen). Her work focuses on reconstructing ancient plant communities and climates. This helps her husband (Dr. Philip J. Currie) to understand the environments that dinosaurs lived in. Dr. Koppelhus's many field projects include dinosaur sites in Alberta, Antarctica, Argentina, and Asia.

**JESSICA EDWARDS** is a graduate of the University of Alberta's Evolutionary Biology program. While at the University of Alberta, she studied the microfossils of the Danek Bonebed and was a teaching assistant for the online course *Dino 101* in its first year.

Made in the USA
Las Vegas, NV
01 September 2022

54518828R00105